THE SEARCH *for* INFINITY

solving the mysteries of the Universe

THE SEARCH *for* INFINITY

solving the mysteries of the Universe

GORDON FRASER
EGIL LILLESTØL
INGE SELLEVÅG

INTRODUCTION BY
STEPHEN HAWKING

LONDON NEW YORK SYDNEY TORONTO

Executive Editor	Robin Rees
Commissioning Editor	William Hemsley
Project Editors	Richard Brzezinski, Chris Cooper
Proofreader	Sean Connolly
Executive Art Editor	Vivienne Brar
Picture Research	Caroline Hensman
Production	Chris Latcham

This edition published 1994 by BCA
by arrangement with Reed Consumer Books Limited

CN 8041

Reproduction by Alphabetset, London
Produced by Mandarin Offset, Hong Kong
Printed and bound in Hong Kong

Contents

The idea for this book came from a series of articles written by Inge Sellevåg, in collaboration with Professor Egil Lillestøl, which appeared in the Norwegian newspaper *Bergens Tidende*, and was subsequently adapted into a booklet *Atomer og kvarker* (Atoms and Quarks). The material was further developed in collaboration with Gordon Fraser at CERN, the European Laboratory for Particle Physics, Geneva, Switzerland.

We thank the Norwegian Research Council, *Bergens Tidende* and CERN for their continued support and assistance, and Robin Rees and his enthusiastic team at Mitchell Beazley who transformed our pile of text into such an attractive book.

Introduction

The human race has always wanted to look beyond the horizon, to see what is out there. In earlier times people thought the sky was a great arch or pudding basin above us over which the Sun and Moon gods drove their chariots once a day. What lay beyond the sky was the realm of the gods, but the part that humans could see was bounded and finite. However, the improvements in astronomical observations that were initiated by Galileo and others in the sixteenth and seventeenth centuries have pushed the limits of our knowledge in turn to the edge of the Solar System, then to the nearby stars and finally to the edge of the observable Universe. Here we are really up against a barrier. For although the Universe may well be infinite in spatial extent, it is definitely finite in time, in the past direction at least, having begun about 15 billion years ago in the Big Bang. This means that we can't see objects more than 15 billion light years away, because the light wouldn't have had time to get here. Thus our horizons are not quite infinite, but they are a hundred thousand billion billion miles away, which is nearly infinite.

At the same time as we were reaching further and further out into space, we were also probing matter to smaller and smaller distances. The invention of the microscope allowed us to see objects like cells and bacteria that are a thousandth of a centimetre across or less. But if one wants to look on smaller scales, one comes up against the difficulty that visible light is made up out of waves whose wavelength, or distances from crest to crest, is between four and eight hundred-thousandths of a centimetre. This means that one can not directly "see" structures smaller than this. To probe more deeply, we have to use something with a short wavelength, Because of a law discovered by the German physicist Max Planck, this means higher energy. That is why giant machines have been built to accelerate particles to enormous energies. By colliding these particles with fixed targets or with similar particles coming in the other direction, we have been able to investigate the structure of matter down to a scale of about a millionth of a billionth of a centimetre. One might think that one could go on building more and more powerful particle accelerators and discover new layers of structure on smaller and smaller scales. But it seems that there's a limit we can not go beyond. A particle with a wavelength shorter than what is called the Planck length, a millionth of a billion-billion-billionth of a centimetre, would have an energy so high that it would form a black hole and fall inside. The particle energy required to reach down to the Planck length is almost a billion billion times the energy attained in the most powerful particle accelerator today. So we are not likely to reach it in the near future, given the present financial situation. But the fact that there is a limiting length, even though very small, is very important. It means that the range of length scales we have to investigate is not quite infinite. Rather it is bounded below by the Planck length, and above by the size of the observable Universe. The radius of the Universe is about 1 with 60 zeros after it times the Planck length. Ordinary, human scales of centimetres and metres come roughly half way in this enormous range between the two limits. That is to say we are very approximately as many times larger than the Planck length as the radius of the Universe is bigger than our size.

On either side of us, the Universe has structure on scales up to about a thousand billion billion billion times bigger or smaller than our own. Because this range is not quite infinite, there is hope that we may one day completely understand the structures of the Universe, from the very smallest to the largest we can know. This book describes the remarkable progress we have already made towards this complete picture. Strictly speaking maybe the title should be *The Search For The Almost Infinite.* The only thing that seems to be unbounded is the power of reason.

Stephen Hawking
Cambridge
February 3rd, 1994

PART ONE
Looking in

Worlds, big and small

POWERS OF TEN

Twentieth-century physics has ventured on two major voyages of discovery. One of these has looked outwards through telescopes towards the edge of the Universe; the other has probed inwards to the microcosmos – the minuscule world of atoms and subatomic particles.

We can see into outer space and sense that there is a great cosmos extending far beyond the faintest visible stars. It is more difficult to grasp that there is a microworld of comparable depth and complexity. The smallest distances visible to the naked eye – for example, the breadth of a hair – are less than a millimetre (a twenty-fifth of an inch). Beyond that everything turns fuzzy.

The eye picks up light and transforms it into signals that the brain interprets as pictures. However, even the sharpest eye cannot distinguish an object smaller than the distance between the sensitive cells on the retina. To see such objects requires a magnifying glass or a microscope.

The first microscope was developed by Anton van Leeuwenhoek, a Dutch clothes salesman with no scientific training whose hobby was making lenses. He built a modest instrument capable of magnifying 200 times. While studying a raindrop in 1683 he discovered the first micro-organisms, which he called animalcules or "little animals". Van Leeuwenhoek soon discovered a "zoo" of tiny organisms. The scientific world was sceptical of his claims even though they were carefully documented. In those days the cheese mite was declared God's smallest creature. From Leeuwenhoek's drawings it was later concluded that he was probably the first to see bacteria, the smallest self-contained living organisms.

Bacteria, typically 1 micrometre (a hundred-thousandth of an inch) across, are the smallest common objects visible with an optical microscope. A micrometre (sometimes abbreviated to micron) is a millionth of a metre, or 0.000001 metre.

There is a method to avoid having to write all these zeros when using very large or very small numbers. A micrometre is a metre divided by 10 six times over, and can be written 10^{-6} metre (said in words as "ten to the power of minus six" or just "ten to the minus six"). The same technique goes for writing large numbers, but this time the "power" is a positive instead of a negative number. Thus 100 million (100,000,000) is 1 multiplied by 10 eight times, and can be written as 10^{8} ("ten to the eight"). If you look at the two examples, you can see that the power is how many zeros there are in the number. So, for example, 10^{32} is 1 followed by 32 zeros – a very large number indeed.

SMALL WORLD

Beyond bacteria, objects are smaller than the wavelength of visible light, and light waves just surge unaffected past them. But light is not the only option. Particles can behave like waves – for example, electrons make waves with wavelengths thousands of times shorter than those of light.

The first electron microscope was developed in 1931 and opened up the world of viruses (10^{-7} metre). Today's electron microscopes can "see" the structures of molecules (10^{-9} metre) and scan the surface of individual atoms. An atom is typically 10^{-10} metres across – 100 million million would cover only a square millimetre (a hundredth of a square inch).

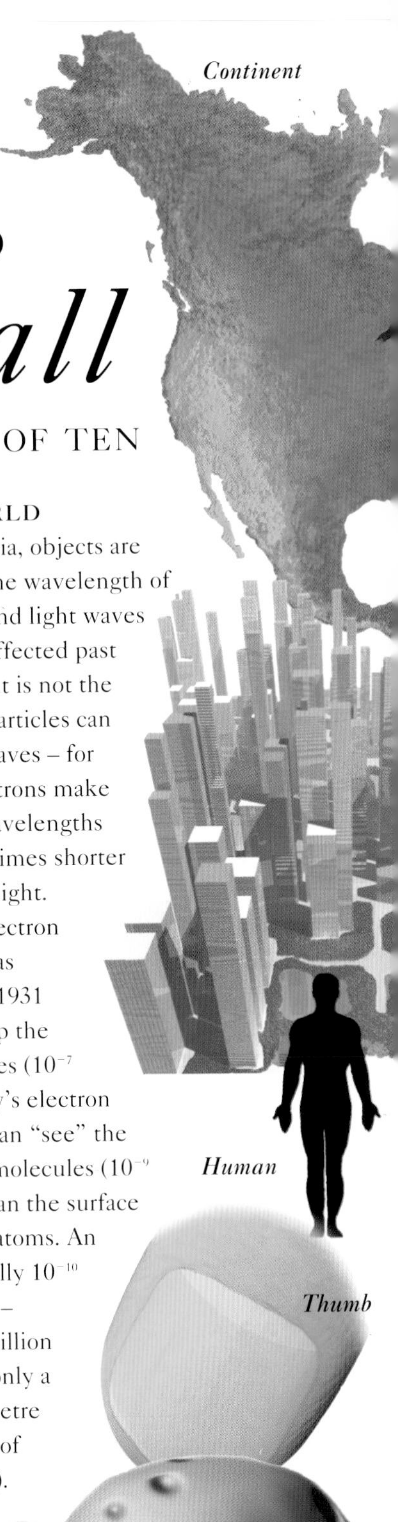

Continent

Human

Thumb

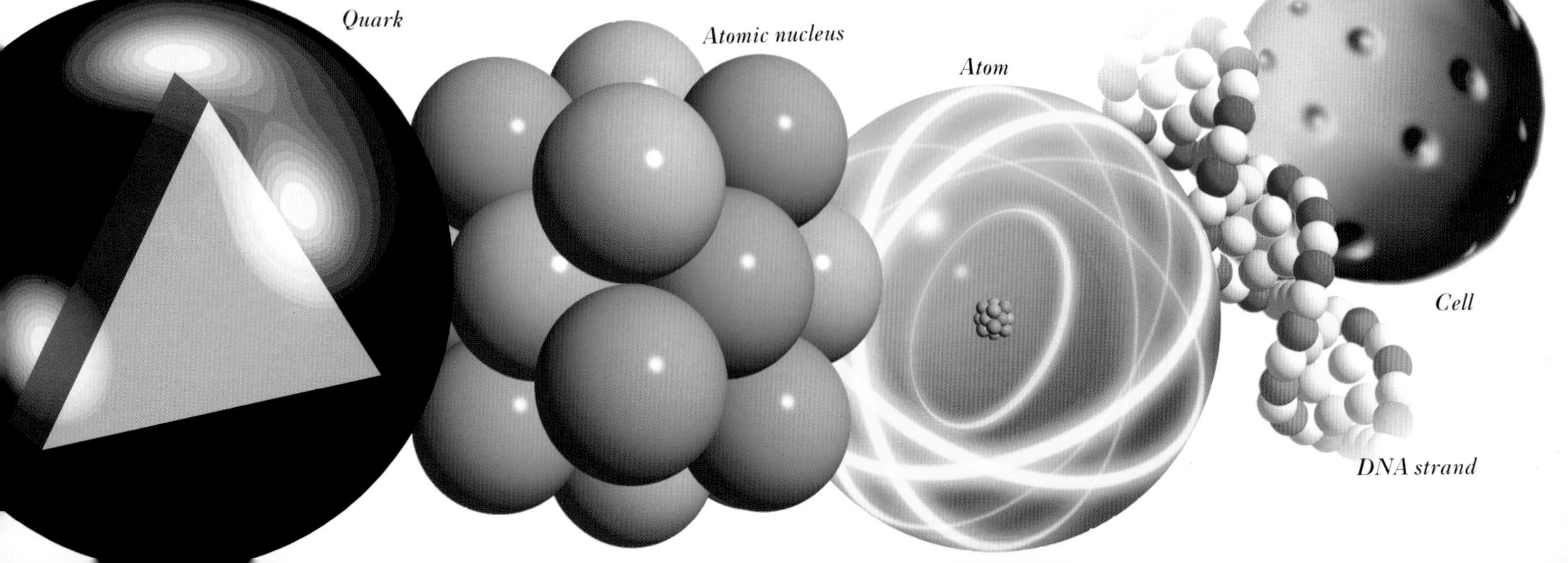

Cell

DNA strand

Atom

Atomic nucleus

Quark

It seems at first paradoxical that something as small as an atom has a vast interior. Deep inside the atom is a tiny dense nucleus (10^{-14} metres across) surrounded by a cloud of electrons that is 10,000 times bigger than the nucleus but carries only one twentieth of one per cent of the atom's mass. Here begins the true microcosmos (a Greek word meaning "little world"). To peer into the microcosmos, physicists use a different kind of microscope, creating still smaller wavelengths using machines called particle accelerators.

Sixty years ago physicists knew that the atomic nucleus consists of protons and neutrons and believed this was the ultimate microcosmos. But studies of cosmic rays from outer space and experiments with more powerful accelerators uncovered even deeper layers of matter. Protons and neutrons are built of smaller particles called quarks. These, together with the electrons, are the pebbles of modern physics.

Levels of matter ***The natural world spans an incredibly large range of dimensions, from the tiny constituents hidden deep inside the protons and neutrons of atomic nuclei through to the threadlike strings in which galaxies cling together. The limited world of human experience roughly midway between these two cosmic extremes.***

Quarks and electrons are less than 10^{-18} metre across. However, the microworld could stretch far beyond this level. New ideas say that particles are not points but little strings 10^{-33} metre long. These minuscule distances are comparable to the size of the Universe in the first fractions of a second after it was born in the Big Bang – the structure of the microcosmos goes back to the ultimate beginning and helps reveal the secrets of the great cosmos.

BIG WORLD

The sky above us is a vast tapestry of twinkling points of light – stars. On a clear night the naked eye can see about 3,000 of them, all very near by cosmic standards, just the nearest edge of the rest of the Universe. We live on a small planet orbiting an average-sized star, the Sun, on the outskirts of a giant spiralling star system – the Milky Way galaxy. This galaxy contains 100,000 million stars, as many stars as it would take grains of sand to fill a large room. Elsewhere in space there are at least 100,000 million other galaxies.

The distances in space are so vast that ordinary distance units no longer make sense. Instead distances are measured using light as a yardstick. The distance travelled by light in empty space in a year is called a light-year and is equivalent to nearly 10 million million kilometres (10^{16} metres, or 6 million million miles). Light takes about four years to travel from Alpha Centauri, the nearest star to us outside our solar system; the well-known Pleiades (Seven Sisters) lie about 400 light-years from Earth; and the distance to our neighbouring galaxy Andromeda, the most remote object familiar to the naked eye, is 2 million light-years.

Because light takes so long to reach us, looking out into space means looking back in time. The starlight we see today was emitted a long time ago. With powerful telescopes, astronomers can look almost to the edge of the observable Universe where they have detected some extremely bright "quasi-stellar" objects, the mysterious "quasars", each blasting out hundreds of times the luminous energy of ordinary galaxies, and more than 10,000 million light-years away. That brings us almost back to the time the Universe was born.

The two long journeys, one going back in time by traversing the increasingly large distances of outer space, the other delving deep into "inner space", probing the micro-mechanisms which governed the early Universe, eventually converge.

The Greek revolution

GREEK PHILOSOPHY AND THE ATOM

About 2,500 years ago, the early Greek philosophers challenged ancient mythology and its explanations of the world by beginning to demand a logical answer to the question "What is the world made of?". This revolution in human thinking led to the birth of natural science.

Greek philosophy is often considered to begin with Thales, who lived in the town of Miletus in Asia Minor around 600 BC. Trying to explain the pattern of the world around him by a fundamental principle, he assumed there was a "primary matter" from which all other substances were made. Observing the importance of water in nature, he concluded that everything was ultimately made of water.

The idea was not entirely new. Several of the ancient creation myths saw water as the primeval substance. But Thales' approach was radically different. Instead of invoking gods and supernatural explanations, he was curious about nature itself and made the first attempts at a logical explanation.

THE DARK ONE

Thales' water theory was controversial. Soon afterwards Anaximander suggested the world was made of an abstract stuff he called "*to apeiron*" (the undetermined). Anaximenes, his pupil, claimed that everything was made of air.

A generation later Pythagoras, famous for his theorem on right-angled triangles, discovered the musical intervals, which rest on simple numerical relationships. On this basis he developed a theory that everything consists of numbers. Numbers are the fundamental entities in the Universe, he said, and considered ten to be the perfect number. By drawing a pentagram – a pentagon with all its diagonals – he discovered the "golden ratio", a proportion claimed to be the most pleasing to the eye. Pythagoras was among the first to be aware of the importance of mathematics in physical science.

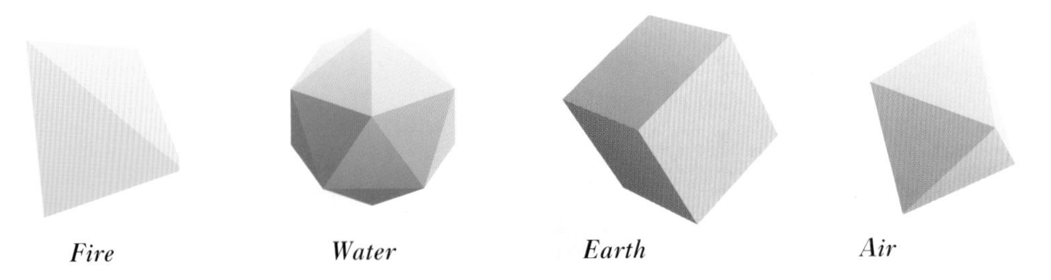

The Four Elements (left) *Many of the ancient Greek philosophers believed that all matter was made up from combinations of four natural "elements". Plato identified each of these elements with a regular geometric solid, although his idea did not become widely accepted.*

Nature explained (below) *The idea of the Four Elements, as shown here in an allegorical Flemish painting, was inspired by natural events, such as storms and volcanic eruptions. It was believed that the elements combined to allow the abundant growth of all plants and animals on Earth. The Four Elements remained the dominant explanation of Nature's processes until as late as the 17th century.*

DEMOCRITUS – A CHEERFUL THINKER

In the fourth century BC, Aristotle wrote of Democritus that "He seems to have thought about everything". As well as his ideas about atoms, Democritus covered a wide range of other subjects – psychology and human nature, agriculture, poetry, geometry, diet, warfare and correct pronunciation. Of the 200 books he is said to have written, unfortunately only a few scattered sentences remain.

A cheerful man, he was known as "the laughing philosopher" and lived to be 90, some say even 110. A native of Abdera in Thrace, he inherited a large fortune and spent it on travelling. One story says he died penniless, another that he was considered mad. The famous Hippocrates was sent to cure him, but returned saying he had never encountered anyone more sane!

CRISIS OF CHANGE

Heraclitus, 40 years Pythagoras' junior and known as "the Dark One" because he spoke in obscure riddles and deliberately made his ideas difficult to understand, said that the world was made of fire. Werner Heisenberg, one of the leading figures of twentieth-century physics, wrote in his book *Physics and Philosophy*, "If we replace the word 'fire' by 'energy' we can almost repeat Heraclitus' statements word for word... Energy is the substance from which all things are made and energy is that which moves."

Heraclitus was primarily concerned with the problem of motion and change. The world is in constant change, he said, describing it as an everlasting fire. Parmenides, a contemporary of Heraclitus, on the contrary claimed that nothing changes. Change is logically impossible, he argued, because it implies a state of nothing or "non-being". Change was merely an illusion.

He started the Greeks on a new track of abstract thinking, without reference to the external world, and produced such a strong argument for his seemingly absurd suggestion that it created a crisis in Greek philosophy. Before further progress could be made, a solution to the problem of change had to be found.

ATOMS AND A VOID

During the fifth century BC, various solutions were proposed. The first was suggested by Empedocles. He argued that the world was made of four basic substances, which he first called "roots" and then later "elements": earth, fire, air and water. These combined to form new substances, and so produced change. Empedocles was a religious mystic and imagined that the elements were bound together and separated by moral forces – Love and Strife.

The next solution came from Anaxagoras, who claimed that all things were made up of indivisible seeds. There is something of everything in everything, he said. Both Anaxagoras and Empedocles believed that matter was continuous, so that in principle it could be divided into ever smaller parts.

Democritus put forward the idea of discontinuous matter. After a certain number of divisions, he said, a limit is reached where the parts cannot be divided any further; all substances are built from invisible and uncuttable particles, which he called "atoms" – from the Greek for "uncuttable". These particles had different shapes, forms and weights, and combined into new substances by moving about in a void or emptiness.

Atoms with God on their side

THE BEGINNINGS OF MODERN SCIENCE

The atomic theory was revived in the seventeenth century, with the birth of modern science. To be accepted, this apparently atheistic theory first had to be modified to incorporate the idea of God. Alternative ideas were put forward to replace the four elements of the Greeks.

Most Greek philosophers rejected atomism, mainly because it contained the idea of nothing (the void). They preferred the theory of the Four Elements, supported by Plato and Aristotle, the giants of Greek natural philosophy. Aristotle also introduced a fifth element, representing purity – "the quintessence" – from which celestial bodies were made. Because of his authority and renown, the element picture won the debate and reigned for 2,000 years, while the atomic picture lay almost forgotten.

However, Democritus' atomic theory was briefly revived by Epicurus around the end of the fourth century BC. He incorporated atomism into a materialistic philosophy aimed at abolishing the fear of gods and promoting happiness and peace of mind. Later, the first-century BC Roman philosopher Lucretius popularized atomism in his book *De Rerum Natura* (On the Nature of Things). Through this book the ideas of indivisible "building blocks" were rediscovered in the seventeenth century.

PIERRE GASSENDI – A THEORY OF GRAVITY

Pierre Gassendi not only "reinvented" atoms but also developed the first concept of gravity. He assumed that each particle in a falling body was pulled downwards by thin strings linked to the Earth. A large object was heavier because it contained more particles and had more strings.

Gassendi speculated that the attraction might extend into space, all the way to the planets, but failed to see how it might explain celestial motion.

Gassendi was born of poor parents at Champtercier in Provence, France, and when aged 40 he became the provost of the cathedral church at Digne.

Isaac Newton in 1687 gave the first valid description of gravity, proposing it as a mutual attraction between all bodies because of their mass, explaining both how bodies fall to earth and how the solar system holds together. He spoke favourably of Gassendi's work and could have been influenced by it.

Also by the seventeenth century scientists began to suspect that the Four Elements were not the whole story. Their scepticism shook the Aristotelian world-view and gave atomism a new chance. But the original atomic theory showed no need for a creator, and no scientific theory could afford to overlook God.

Atomism was revived by the French Catholic priest and philosopher Pierre Gassendi. He studied the work of Epicurus and set out to purge its atheistic implications. God created all matter, Gassendi said, but the actual behaviour of matter was due to atoms and their motion in the void. In 1624 Gassendi wrote a book entitled *Excercitationes Paradoxicae Adversus Aristotelos*, which attacked Aristotle. But in France anti-Aristotelian teaching was forbidden and Gassendi's ideas in physics were not published in full until 1658, three years after his death.

The alchemist at work ***Often thought of as merely the quest to turn lead into gold, alchemy was the first genuine attempt to investigate the nature of matter by experiment. The alchemists hoped to prove Aristotle's Four Elements theory, and in the process of their work invented many techniques that are still used in chemistry today.***

THE MAD DUCHESS

Atomism gained ground in the 1640s, but was still controversial. Margaret Cavendish, Duchess of Newcastle, an eccentric English lady who earned the name "the mad Duchess", both embarrassed friends of atomism and shocked its enemies. She was a poet, biographer and actress, and also wrote several widely read books on natural philosophy.

In 1653 Lady Margaret published two volumes of verse she called "fancies". Here she made the almost heretical claim that atoms by their own motion made up the world. "It is better to be an atheist than superstitious", she wrote.

The Duchess of Newcastle counted four types of atoms – square, long, round and sharp. She applied her atomic ideas also to medicine and psychology. Sickness, she said, was caused by atoms "fighting"; while memory was "atoms in the brain set on fire".

THE SCEPTICAL CHEMIST

Robert Boyle, an Irish physicist and chemist who was a contemporary of Margaret Cavendish, distanced himself from what he called the "modern admirers of Epicurus who banish God from the Universe". Boyle was a deeply religious man, and he never mentioned the name of God without a short pause for veneration. He avoided using the term "atomism" because of its atheistic overtones, preferring to refer to such ideas as "corpuscular philosophy".

Boyle made atoms respectable by incorporating them as an integral part of chemistry. Matter, he claimed, existed as "*prima materia*", indivisible and impenetrable, endowed with motion by God. In his book *The Sceptical Chymist* he did not accept the old idea of the Four Elements. He mocked the alchemists, who were striving to turn base metals into gold as the ultimate proof that the Four-Elements theory was right.

A true element, Boyle said, cannot be decomposed into simpler substances. He recognized that air is not homogeneous, but in those days chemical analysis was rudimentary. It took another hundred years before chemists could build on Boyle's new definition and show that the Four Elements were in fact composite substances made up from elements. In his studies of gases, Boyle also showed for the first time that air has weight.

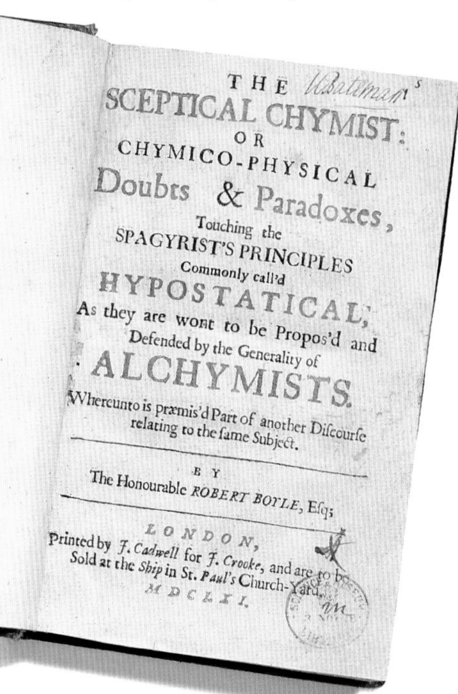

THE SCEPTICAL CHYMIST: OR CHYMICO-PHYSICAL Doubts & Paradoxes, Touching the SPAGYRIST'S PRINCIPLES Commonly call'd HYPOSTATICAL; As they are wont to be Propos'd and Defended by the Generality of ALCHYMISTS. Whereunto is præmis'd Part of another Discourse relating to the same Subject.

BY The Honourable ROBERT BOYLE, Esq;

LONDON, Printed by J. Cadwell for J. Crooke, and are to be Sold at the Ship in St. Paul's Church-Yard. MDCLXI.

The title page of The Sceptical Chymist ***Published in 1661, Boyle's book laid the foundation for modern scientific method by insisting that all theories were invalid unless confirmed by experiment. It has been seen as the first truly scientific work.***

Particles or waves?

TWO RIVAL THEORIES OF LIGHT

The nature of light has been one of science's oldest preoccupations. In the seventeenth century physicists pioneering the new scientific revolution had two theories – one that light consists of particles, the other that light is waves. The wave idea quickly became dominant.

The early Greek philosophers believed that light was emitted by the eye and produced sight when it hit objects. Since light cast sharp shadows, it clearly travelled in straight lines. The Greeks were familiar with the bending of light when passing from air through a denser medium like water or glass (refraction) and noticed that smooth surfaces reflect light. They also experimented with magnification by filling glass bulbs with water.

Later it was recognized that the eye receives light, and Aristotle interpreted visibility as illuminated space. This view was itself overthrown when scientists in Bologna discovered a rock that continued to glow faintly in the dark after it had been exposed to sunlight. The discovery made them realize for the first time that something does not have to be illuminated by the eye to be visible. They called this rock with luminous memory the "solar sponge". Galileo Galilei brought some fragments of the rock with him when he travelled to Rome in 1611 to demonstrate his first telescope.

Galileo concluded that light consists of particles. The solar sponge, he said, attracts light particles as a magnet attracts iron filings. Galileo initially used the word "atom" only to describe light particles. He referred to matter particles as "the smallest quanta" (minimi quanti). Later, however, he abandoned both his atomistic views and his particle theory of light.

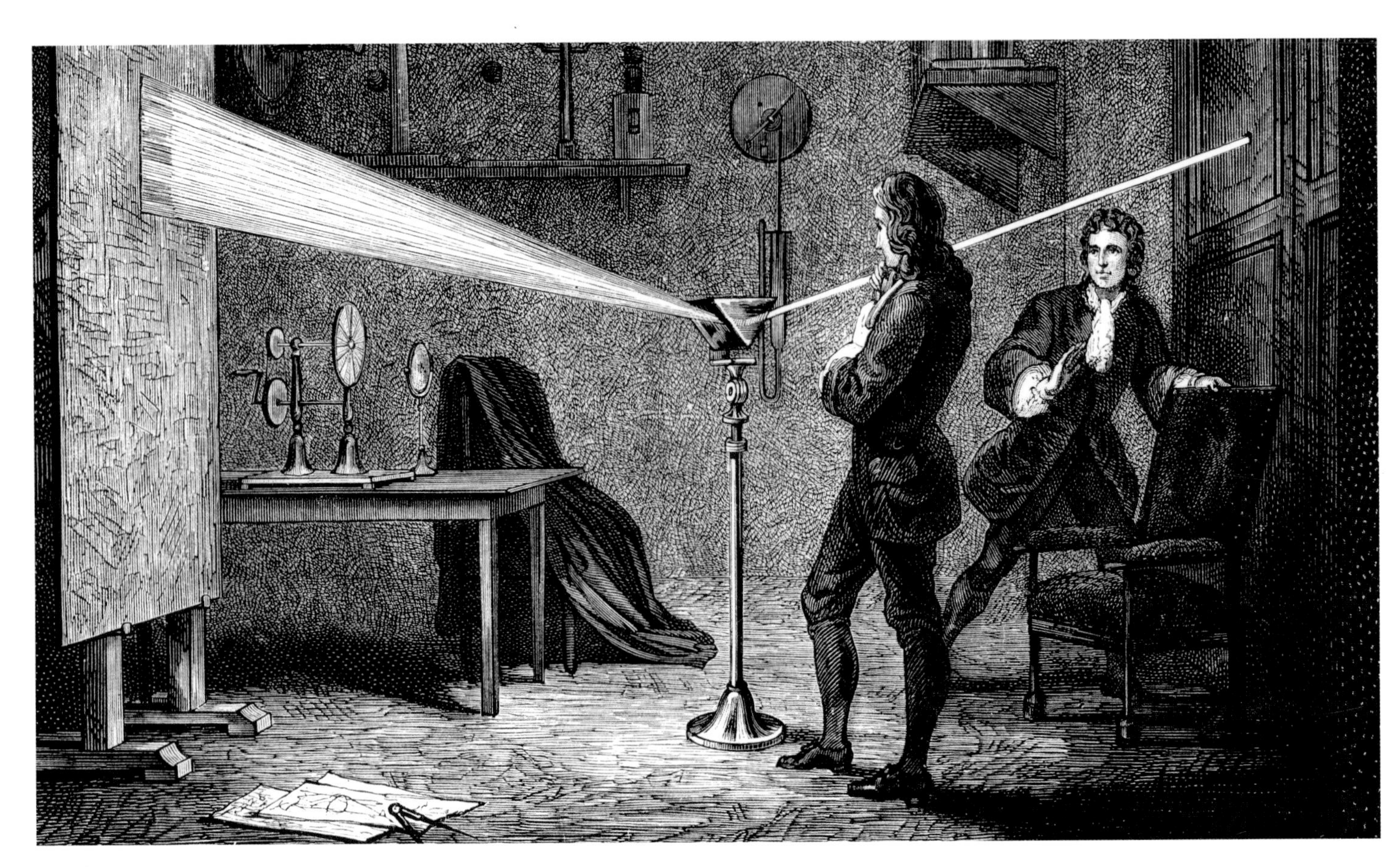

Isaac Newton's spectrum *A 17th-century print showing Newton at work. In 1666, he intercepted a sunbeam with a prism, which split the light into a band of colours, which Newton called a "spectrum", a Latin word meaning "appearance". Three years later, he used a lens to reconstitute white light from the spectrum, showing that white light is a mixture of rays.*

LIGHT PARTICLES

In England, Isaac Newton carried out the first carefully planned scientific experiments on light. He split sunlight with a prism, and found it was composed of a "spectrum" of colours. He also studied the beautiful changing display of colours in soap bubbles.

Newton proposed that every material object is a combination of matter and a luminous substance which emits light. Light itself, he thought, consisted of tiny particles (corpuscles) that were travelling at enormous speed. It is reflected by a mirror because the particles bounce back from the surface, like a ball hitting a wall, and is bent on entering a refracting medium such as water or glass, because the speed of the particles is altered. He believed that space was also filled with light particles. In this way Newton reintroduced the classical concept of the ether, the invisible medium that, according to Aristotle, filled the heavens. Newton's treatise on *Opticks* was finally published in 1704.

AT THE SPEED OF LIGHT

The Danish astronomer Ole Rømer (shown here in his Copenhagen observatory) was the first to prove that light has a finite speed. In 1675 he noticed that Jupiter's moon Io went into eclipse slightly earlier when Jupiter was at its closest to Earth. He deduced that the difference was due to the time it takes light to travel across the changing distance, and he predicted, correctly, the times of future eclipses. His observations were later used to calculate that light travels at 225,000 kilometres (140,000 miles) per second. The correct value is about 300,000 kilometres (186,000 miles) per second.

WAVE THEORY

In 1690 Christian Huygens, a Dutch physicist who invented the first pendulum clock, suggested that light is made up of waves that propagate the same way as sound: "When we see a luminous object, it cannot be by any transport of matter coming to us from this object, in the way in which a shot or an arrow traverses the air". If light consisted of particles, he said, rays from different directions would collide.

Light is transmitted from a source to the observer "by the movement of matter which exists between us and the luminous body", Huygens said, falling back on the idea of an ether filling all space. The main objection to his theory was that light appears to travel in a straight line, rather than diffusing round corners, as an ether picture would suggest.

For about a century the two theories of light competed with each other. Newton's particle theory was by far the more popular, largely because of Newton's fame and scientific reputation, and because the theory gave the most logical explanation for refraction and diffraction.

Opinion changed dramatically in 1803 with the English physicist Thomas Young's two-slit experiment, which seemed to exclude any possibility that light was particles. The wave theory ruled supreme through the nineteenth century. Later, quantum theory readmitted the particle explanation (see page 30). Light is in fact both particles and waves.

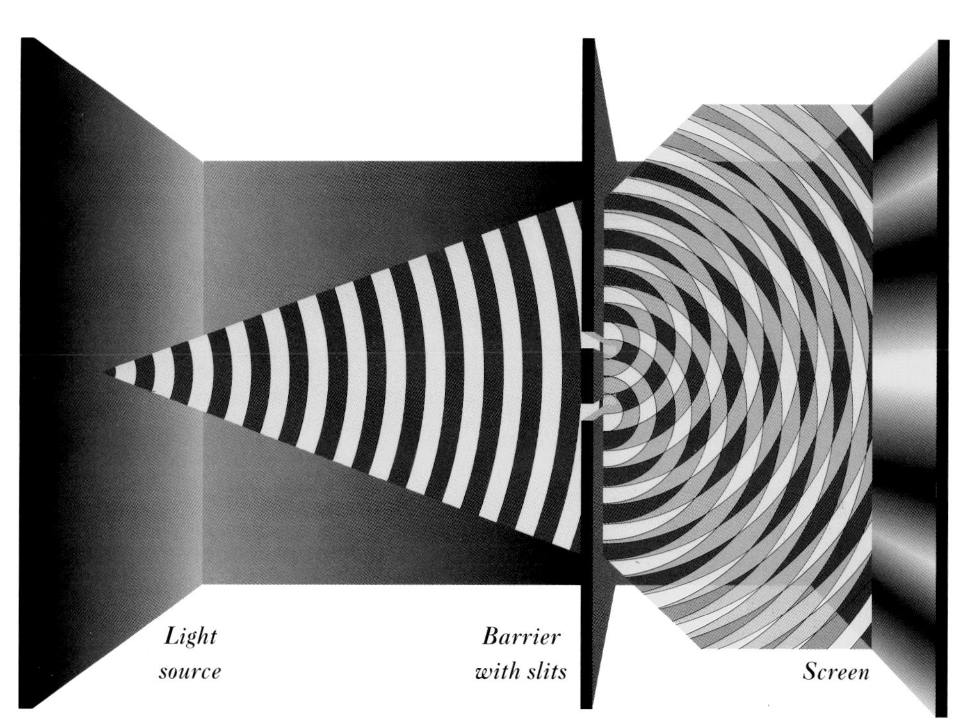

Young's two-slit experiment *Like sea waves that battle with each other, light rays sometimes cancel out and sometimes get reinforced. Young passed light through two slits close together in a barrier, and the waves of the two resulting beams joined together producing crests and troughs, which showed up on a screen as a so-called interference pattern. This seemed to prove conclusively that light was made up of waves not particles.*

Confusion in chemistry

PERIODICITY OF THE CHEMICAL ELEMENTS

With the Greek idea of the Four Elements abandoned, scientists quickly began to identify the modern chemical elements. By the nineteenth century, the mounting number of elements, now thought to be built from atoms, confused chemists. The hunt began for an underlying pattern.

In the 1780s the French chemist Antoine Laurent Lavoisier paved the way for a scientific atomic theory by using balances and careful measurement. He showed how compounds were made up of two or more elements, and formulated an important law, saying that the total amount of matter is the same both before and after a chemical reaction.

Lavoisier was an anti-atomist. He advocated a pragmatic definition of an element and said that further discussion on the nature of matter was metaphysics and not real science. However another French scientist, Joseph Gay-Lussac, discovered that when hydrogen and oxygen formed water, they always did so in fixed proportions. This signalled that elements contained some basic units.

Lavoisier's apparatus for separating water (above) *Water drips through a red-hot iron gun-barrel. Iron oxide forms inside the barrel; hydrogen gas is given off and collects on the right. This proved that the Aristotelian "element" of water was made up of two more fundamental elements.*

JOHN DALTON – THE WEATHER MAN

John Dalton (1766–1844), was born in the English Lake District and brought up as a Quaker. He spent most of his life in Manchester, earning his living as a private tutor. A clumsy ill-mannered man, he was generally unprepossessing and often incomprehensible.

Dalton's passion was meteorology. Throughout his life he kept a record of the local weather, and his first scientific work was a book of meteorological observations. His discoveries in atomic theory were prompted by his interest in air and water vapour. His investigations led him to analyse many gases, which he collected himself, as in this painting by Ford Maddox Brown, *Dalton Collecting Marsh Fire Gas.*

Dimitri Ivanovich Mendeleyev (1834–1907; above right) *While professor at the University of St. Petersburg, he proposed a new way of classifying the known elements. His "periodic table" still forms the roadmap of modern chemistry.*

Dalton's list of the elements (right) *Dalton made a serious attempt at measuring the atomic weights of the elements. His list of 1806–7 contained many errors, but still proved to be a useful tool for chemists.*

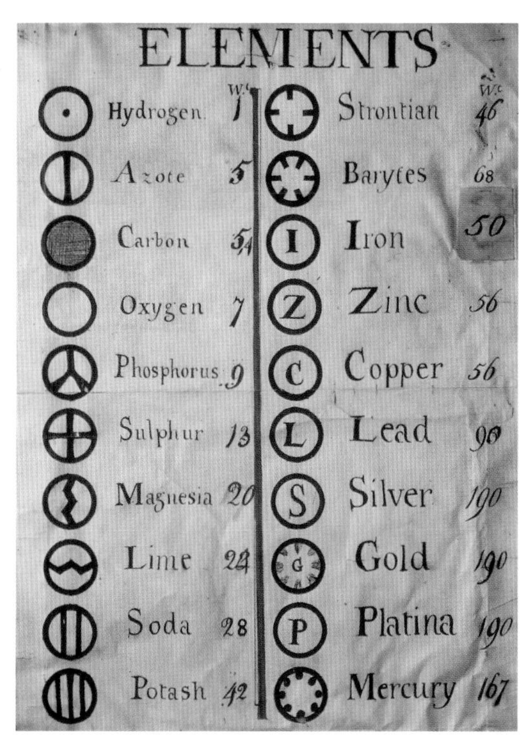

ELEMENTS

Element	Wt	Element	Wt
Hydrogen	1	Strontian	46
Azote	5	Barytes	68
Carbon	5,4	Iron	50
Oxygen	7	Zinc	56
Phosphorus	9	Copper	56
Sulphur	13	Lead	90
Magnesia	20	Silver	190
Lime	24	Gold	190
Soda	28	Platina	190
Potash	42	Mercury	167

In 1803 the Englishman John Dalton built on the basic-unit idea. He suggested that each different element had its own type of atom, and that all atoms of a given element were identical. Dalton saw atoms as hard and indestructible, rather like billiard balls. When elements combined, he said, their atoms clustered together into what the Italian Amadeo Avogadro later called molecules. From these ideas it was possible to calculate the weights of the atoms of different elements in relation to each other. Dalton published the first table of such "atomic weights".

Dalton's theory was controversial. Most chemists argued there was no need for atoms, which were probably unobservable and his theory therefore meaningless. A group led by the English chemist Sir Humphrey Davy criticized Dalton because his elements had different atoms. Davy held that all matter was made up of identical atoms, and that the number of real elements was extremely small.

Dalton knew about 20 elements, but the list kept growing with time, and chemists began to suspect there was some kind of order behind this increasing complexity. In particular they were intrigued that some physically very different elements (for example the gas fluorine, the liquid bromine and the solid iodine) neverthe-less had strikingly similar chemistry.

THE PERIODIC SYSTEM

The puzzle was solved by the Russian chemist Dimitri Ivanovich Mendeleyev. In 1869 he devised a new classification system, the "periodic table". Listing the elements in the order of their atomic weight, he found that elements with similar properties appeared at regular intervals (or "periods") and could be lined up in vertical columns.

To make his periodic table work, Mendeleyev had to leave gaps in it, predicting that the missing elements would turn up. Three new elements – germanium, gallium and scandium – were soon found. Their discovery was a major triumph, both for the periodic table and for Dalton's atomic theory.

For a time it looked as if the 2,000-year quest to understand matter was complete. But important questions still remained. What, for instance, gave the elements their neat periodicity? Science had to wait 60 years before Niels Bohr explained these regular properties as being due to repeating patterns in the arrangement of electrons within atoms.

In a lecture at the Royal Society in London in 1889, Mendeleyev said a new mystery of nature had unfolded. He pointed to people's natural reverence for the stars and their patterns, and said: "But when we turn our thoughts towards the nature of the elements and the Periodic Law, we must add the nature of the elementary individuals which we discover everywhere around us. Without them the stellar sky is inconceivable."

Waves across the ether

THE THEORY OF ELECTROMAGNETISM

A flash of lightning and the swing of a compass needle are at first sight very different phenomena. But they have a deep connection. In 1864 James Clerk Maxwell put forward an elegant theory saying that electricity and magnetism have a common underlying force: electromagnetism.

The two forces of electricity and magnetism have been known since ancient times. According to legend, both were first described by the Greek philosopher Thales around 600 BC. He discovered that pieces of iron ore from Magnesia in Asia Minor stuck to each other, and that a piece of amber ("*elektron*" in Greek) when rubbed on his clothing attracted feathers, hairs, dust and other light materials, an effect now known as static electricity.

Research into electricity and magnetism progressed in the eighteenth century. In 1752 Benjamin Franklin showed that lightning is an electrical phenomenon. However, nobody suspected the two forces were related. The first hint came in 1819 when the Danish physicist Hans Kristian Oersted showed that an electric current made a magnetic needle swing violently.

Oersted's discovery attracted attention all over Europe. In England in 1831, Michael Faraday, a bookbinder and self-educated scientist, showed the reverse effect: a moving magnet could produce an electric current. He went on to build the first dynamo, laying the foundations for modern electrical engineering.

To describe how electric and magnetic forces acted, Faraday introduced a new concept called a field. At that time the only other known natural force was gravity, which in 1687 had been described by Isaac Newton as a universal attraction between the mass of objects. Newton saw it as "action at a distance", transmitted instantaneously across space.

Faraday was intrigued by the pattern that appeared when a sheet of paper covered with iron filings was put above a magnet. The filings make a picture of the lines of force. He saw the magnetic force as behaving as if it were some kind of tensioned elastic. A similar effect happened with electric charges, the source of the electric force. Faraday's limited mathematics forced him to explain physics through models.

Early 19th-century electrotherapy (above) *The early interest in electrostatics showed that electricity was a force – it could literally make hair stand on end. Few suspected that electricity was closely related to magnetism, and that both were aspects of a single force: electromagnetism.*

Marconi's first radio station (right) *Hertz's pioneering radio apparatus allowed him to send signals over about 30 m (100 ft). Guglielmo Marconi greatly improved on the technology in experiments at his father's villa near Bologna, Italy, transmitted the first transatlantic radio signal in 1901.*

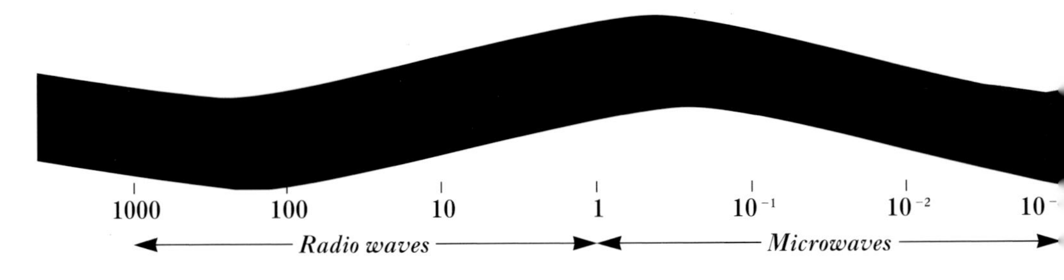

The electromagnetic spectrum *The spectrum extends from the extremes of short wave gamma rays to long radio waves (wavelengths given here in metres). Visible light and infra-red radiation (heat), the only components that pass easily through Earth's atmosphere, support life.*

JAMES CLERK MAXWELL – OVALS AND SATURN'S RINGS

While still at school, James Clerk Maxwell (1831–79) invented a mechanical method for drawing ovals. An article on the method was printed in the *Proceedings of the Royal Society of Edinburgh,* the town where he was born and lived his early life. At the age of 26, already a professor in Aberdeen, he received a prestigious prize for his investigations of the stability of Saturn's rings. From 1860 to 1865 he was professor at King's College, London, but retired to Scotland and a life of seclusion while he wrote his epic 900-page *Treatise on Electricity and Magnetism,* finally published in 1873. In 1871 he became Professor of Physics at Cambridge University's new Cavendish Laboratory.

ELECTROMAGNETIC WAVES

It took the genius of the Scottish physicist James Clerk Maxwell to explore the full consequence of Faraday's discovery. In 1864, Maxwell formulated a set of four elegant equations describing the whole of electricity and magnetism. These suggested that electricity and magnetism were two manifestations of a common force which he called electromagnetism. Maxwell's theory had a startling and completely unexpected bonus. According to the equations, the electromagnetic force travelled through space like a wave and at the speed of light. Maxwell reasoned that light itself must therefore consist of electromagnetic waves. This was an enormous surprise to everyone.

NEW FORMS OF LIGHT

Maxwell boldly predicted that these electromagnetic waves could have a wide range of wavelengths, not just those of light, which are less than a thousandth of a millimetre (400 thousandths of an inch). This was confirmed in 1887, when Heinrich Hertz, Professor of Physics at Karlsruhe in Germany, produced the first radio waves. Radio waves are a form of invisible light with longer wavelengths.

To explain how electromagnetic waves travelled, Maxwell supposed that space was filled with an invisible elastic substance and recalled the old idea of the ether, which Aristotle had assumed filled the heavens. Maxwell said the ether consisted of molecules and described it as "a material substance of a more subtle kind than visible bodies, supposed to exist in those parts of space which are apparently empty".

The nature of the light-carrying ether was hotly debated. Sir George Stokes in England held that it was some kind of jelly. Lord Kelvin claimed it was air-like, so that the Earth's atmosphere extended throughout the Universe. In 1905, however, Einstein's theory of relativity showed that light always travelled at the same speed in a vacuum and finally eliminated the need for an ether.

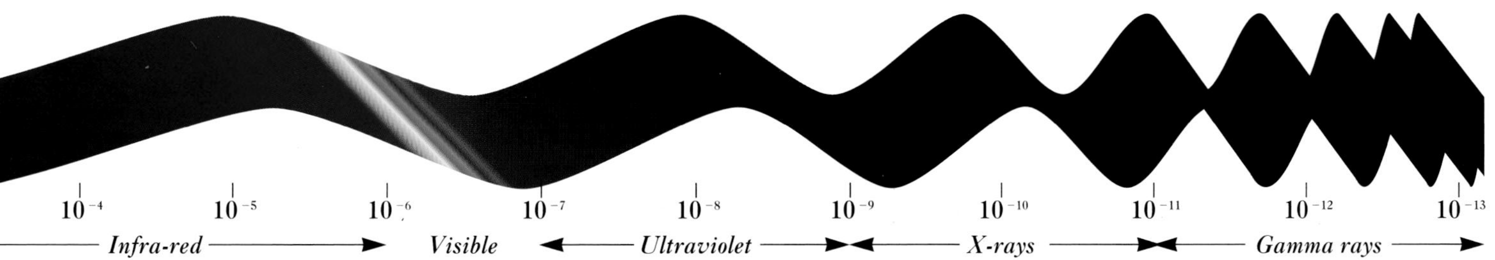

All-revealing radiation

The Discovery of X-Rays

At the end of the nineteenth century many physicists were intrigued by a strange electrical phenomenon called "cathode rays". Their research into these rays began an unexpected chain of events, which culminated with the shattering of the sacred idea of an indivisible atom.

In the 1850s, German physicists tried to send electricity through a vacuum. When two metal plates sealed into the glass at opposite ends of the vacuum tube were connected to a battery, a green glow appeared at the end of the tube by the negative plate, or cathode.

In 1886 Eugene Goldstein called the phenomenon "cathode rays". In 1892 Philipp Lenard found that the rays could pass through a thin aluminium window in the tube, penetrating about 8 centimetres (3 inches) into the surrounding air. He looked at how different substances, including a photographic plate, absorbed the rays if put in their path.

In the afternoon of 8 November 1895, Wilhelm Konrad Röntgen, Professor at the University of Würzburg, prepared an experiment, covering a cathode-ray discharge tube with a layer of thick black paper. He had heard about Lenard's discovery and wanted to see for himself. Not expecting anything spectacular to happen, he darkened the room and switched on the apparatus.

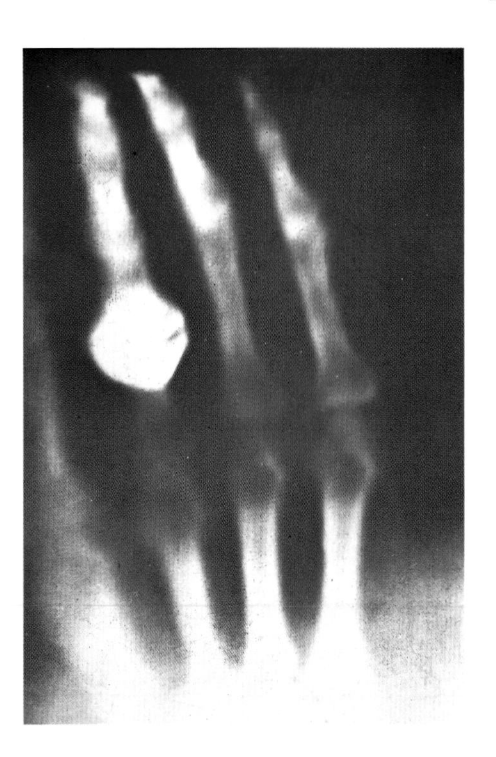

The first X-ray picture (above) *Röntgen made this picture showing the bones in his wife's hand. The bones, and her ring, absorbed the rays, leaving those parts of the film unexposed and therefore clear.*

To Röntgen's surprise a fluorescent screen a full 2 metres (6 feet) from the shrouded tube started to glow. He immediately suspected that he had come across a new kind of radiation, much more penetrating than cathode rays. Over the following weeks he worked feverishly, alone in his laboratory. When his wife asked him what he was doing, he replied that if people found out, they would probably say "Röntgen has gone crazy".

"X" for the unknown

The rays passed not only through paper, but also through a thick book, blocks of wood, metal foils and even human flesh. "If the hand is held between the discharge tube and the screen, the darker shadow of the bones is seen, surrounded by the faint shadow of the hand itself," Röntgen wrote.

Unable to explain what the rays were, he simply called them X-rays ("X" for "unknown"). By Christmas he was ready to announce his discovery. He printed a ten-page report and on New Year's Day 1896 mailed copies to leading physicists. Returning home from the post office he said to his wife "Now the fun can start". His report had included an X-ray picture of his wife's hand, showing the bones of her fingers and her ring. She had been not just surprised but horrified by the picture, associating the skeletal features that it revealed with death.

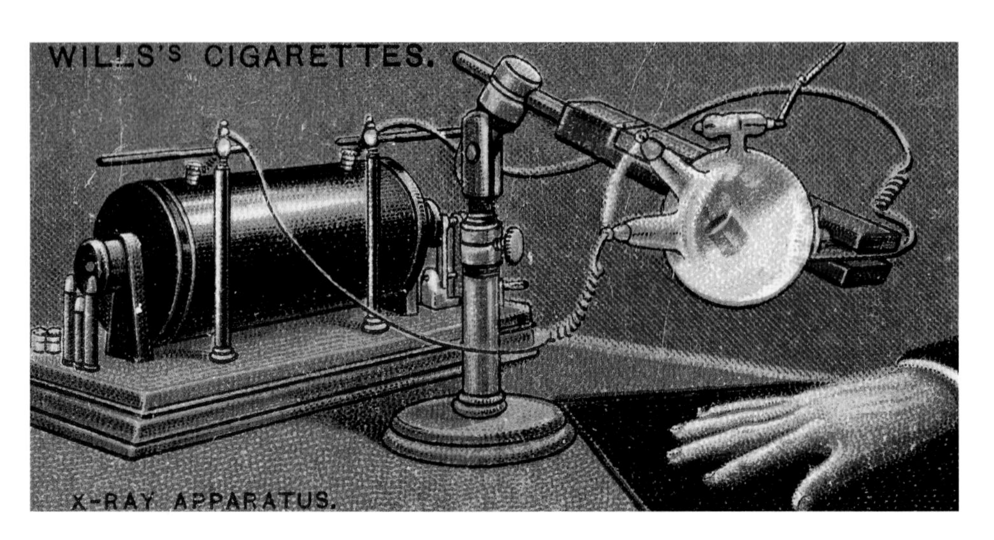

An X-ray apparatus of 1915 *Popular interest in X-rays when they were first announced was enormous, and remained so 20 years after their discovery; this picture appeared in a series of cigarette cards published by the Wills company.*

INVISIBLE PHOTOGRAPHY

The news of X-rays first appeared in a Vienna newspaper on 5 January 1896 and spread like wildfire all over Europe. It was the scientific sensation of the century, producing more than 50 books and papers, and more than 1,000 newspaper articles in that first year alone. Popular journals teased the public with reports of "new all-revealing rays". Women were advised to wear lead-lined clothes to protect themselves from being surreptitiously revealed.

X-rays quickly became a useful tool in medicine. By the spring of 1896, dentists were making X-ray photographs of teeth and doctors were using the new radiation to examine broken bones. There were even efforts to take moving X-ray pictures. Unaware of the dangers of radiation, early experimenters suffered severe "sunburn" and even lethal injuries.

The true nature of X-rays remained a mystery until 1912, when Max von Laue in Germany showed that the rays were refracted – that is, their direction was altered – on passing through a crystal. This showed that X-rays are electromagnetic rays like light, but with a much shorter wavelength. X-rays are more easily absorbed by denser materials containing heavier atoms, so that bone, with its high mineral content, shows up against flesh when X-rays pass through the body. For special X-rays, patients have to swallow or be injected with "tracers" containing heavier atoms, such as barium or iodine.

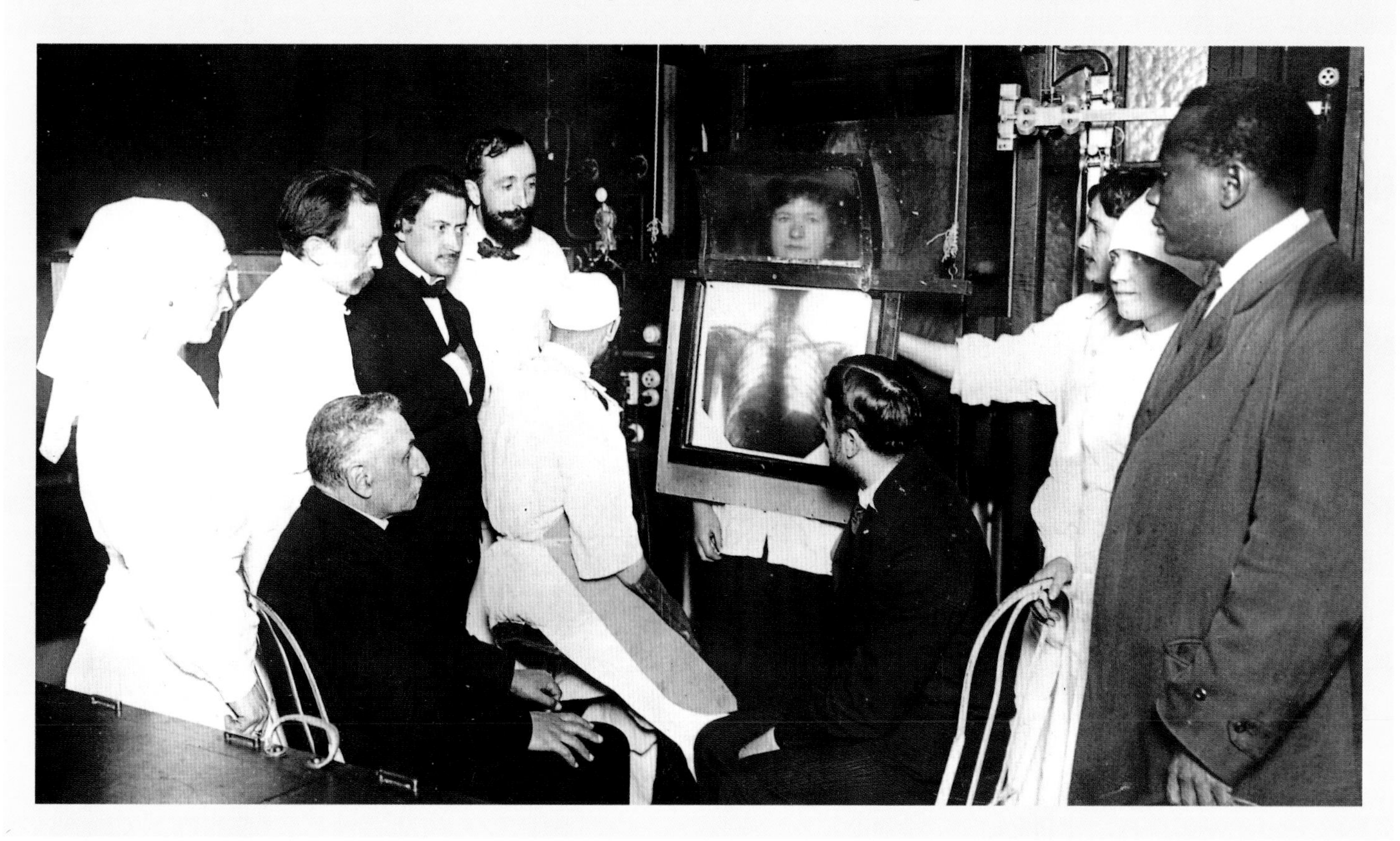

Röntgen, the son of a Rhineland textile merchant, was considered by many to be a wizard after his discovery of X-rays. He was summoned by the German emperor Kaiser Wilhelm to give a private X-ray demonstration. Röntgen regretted the degree of controversy that was surrounding his discovery, claiming that the press had distorted his findings. "In a few days I was disgusted with the whole thing", he said in a rare interview. However, Röntgen was pleased when X-rays went on to become useful in medicine. He did not take out any patents on his discovery, and he did not become involved with the subsequent development of X-ray technology.

The total period of Röntgen's X-ray research covered just 18 months. He published only three papers on the subject, and gave one public lecture, in his home town of Würzburg, on 23 January 1896. In 1901 he was awarded the first Nobel Prize for Physics, but did not give a Nobel lecture. He donated the prize money to the University of Würzburg.

A lifelong friend called Röntgen the epitome of nineteenth-century ideals: a strong and honest man, devoted to his work, and with total integrity. In later years Röntgen avoided meeting his colleagues because of his embarrassment at the publicity X-rays had received.

Opening a new window

THE CURIES AND RADIOACTIVITY

On 20 January 1896, two weeks after Röntgen had announced his discovery of X-rays, the French Academy of Sciences held a special meeting. This gathering was the starting point for the next surprise development in the world of atoms: the discovery of radioactivity.

When the French Academy of Sciences met in Paris on 20 January 1896 to discuss the new X-rays, Professor Henri Becquerel was one of many fascinated scientists in the audience. He was intrigued to learn that the X-rays were produced in the luminous spot where cathode rays hit the glass wall of a high voltage tube.

Becquerel had done research on materials that became luminous after exposure to sunlight and wanted to see if they gave off X-rays as well. He placed some of these materials on photographic plates that were wrapped in opaque black paper, and put them in direct sunlight for several hours. He then developed the plates to see if X-rays were penetrating the black paper. His experiments were going well, but towards the end of February, the sky clouded over. While waiting for the weather to improve, Becquerel shut some partially exposed material and a photographic plate away in a drawer.

On Sunday 1 March the sun shone again. Before continuing the experiment, Becquerel, in his typical systematic manner, decided to develop the plate from the drawer. Although it had been lying in total darkness, he thought there might be some traces of exposure left from the day he interrupted his investigations. To his astonishment the plate had been heavily fogged by invisible rays.

With this new finding, Becquerel's research changed direction. He showed that uranium compounds emitted the rays without needing to be exposed to light or any other source in order to "activate" the uranium. The uranium metal itself was the source of the mysterious radiation, which he called "uranium rays".

Radium craze *A "radioactive-water" vial that once contained a solution of radium-226 and radium-228, and which was still dangerously radioactive after 70 years. Soon after the discovery of radium, reports suggested it might be beneficial to health in small quantities; this led to a boom in radioactive creams and potions. With nobody aware of the hazards, radium became a craze. In 1903 a man suffered severe hallucinations after drinking a radium solution. In San Francisco showgirls wearing radium-painted costumes danced on a darkened stage.*

HEROIC EFFORT

Becquerel's uranium rays were not an immediate sensation like Röntgen's X-rays. They remained a scientific curiosity for two years until Maria Sklodowska, a young Polish woman who came to Paris in 1895 to study chemistry and who married physicist Pierre Curie, chose to study uranium rays for her doctoral thesis.

Wanting to find if other substances emitted the rays, the Curies discovered that the metal thorium was also "radioactive", as they called it. Then they found that pitchblende, the ore from which uranium is extracted, emitted rays even more intense than those of uranium itself.

Working under miserable conditions in a shed, the Curies carefully examined 100 kilograms (220 pounds) of pitchblende residues. In 1898 they identified two new highly radioactive substances, which they called polonium and radium – these were respectively 300 and 2-million times more radioactive than uranium!

SUCCESS AND TRAGEDY

In 1902, after a heroic effort, they managed to isolate a tenth of a gram of pure radium. The metal was luminous, giving out an eerie bluish light. The news of this bizarre new element made radioactivity known to the public. Marie and Pierre Curie shared the 1903 Nobel Prize for Physics with Becquerel and, in 1911, Marie was awarded the Nobel Prize for Chemistry.

After his radioactive discoveries, Pierre Curie became Professor of Physics at the Sorbonne. On 19 April 1906, he died in a Paris street accident, killed instantly under the wheel of a truck. He was probably suffering at the time from radiation

sickness caused by experiments in which he was studying the effects of radioactivity on his own body. Marie succeeded her husband, becoming the first woman professor at the Sorbonne. In 1911 she was the target of lurid newspaper reports accusing her of having an affair with the French physicist Paul Langevin. During World War I, Marie organized special ambulances carrying X-ray equipment to the front, becoming head of a Red Cross radiological unit. She died in 1934 of leukaemia, the result of many years of innocent exposure to radiation.

The Curies *Pierre Curie married Maria Sklodowska in 1895. In 1903 they shared the Nobel Physics Prize with Henri Becquerel. Marie Curie went on to receive the Nobel Chemistry Prize in 1911. The Radium Institute was established in Paris to honour their work on radioactivity.*

URANIUM – A USELESS METAL?

Uranium, the element that revealed radioactivity, was discovered in 1789 and named after the planet Uranus, discovered eight years earlier. Apart from its high density, almost twice that of lead, there seemed to be nothing special about uranium. For a long time it was an oddity at the end of the periodic table with the highest known atomic weight of all the naturally occurring elements; it was known as "the useless metal" as it seemed to have no practical value. Mainly composed of the variety (isotope) uranium-238, natural uranium contains about 0.7 per cent uranium-235, a raw material for atom bombs and nuclear reactors. Uranium is found naturally in yellowish gummite and in black tar-like pitchblende (right), which also contains uranite and traces of radium and polonium.

The first sub-atomic particle

DISCOVERY OF THE ELECTRON

There was controversy at the end of the nineteenth century over the nature of the mysterious "cathode rays" emitted when electricity was passed through a vacuum tube. The debate was settled in 1897 when J. J. Thomson established that the rays were a stream of particles – electrons.

There were two theories about cathode rays, the mysterious glow that appeared when electricity passed through a vacuum tube. Most German physicists believed the rays were a form of electromagnetic radiation, similar to light. British scientists, however, thought them to be particles, because of the sharp shadow cast if an object was put in the rays' path.

The idea that electricity, like matter, consisted of invisible particles had developed in 1833 when Michael Faraday discovered the laws of electrolysis – the chemical changes that are caused by an electric current. This change involved electrically charged atoms, which Faraday called "ions" – from the Greek verb "to go". The Irish physicist George Johnstone Stoney in 1881 introduced the name "electron" for the simplest of the ions.

Sir William Crookes, a flamboyant English chemist who in 1861 had discovered the element thallium, claimed that cathode rays were a new type of matter. He suggested that matter could have a fourth state – which he called "radiant matter" – in addition to the states of solid, liquid and gas. He put forward a theory of matter solidifying from a fiery mist in the early Universe.

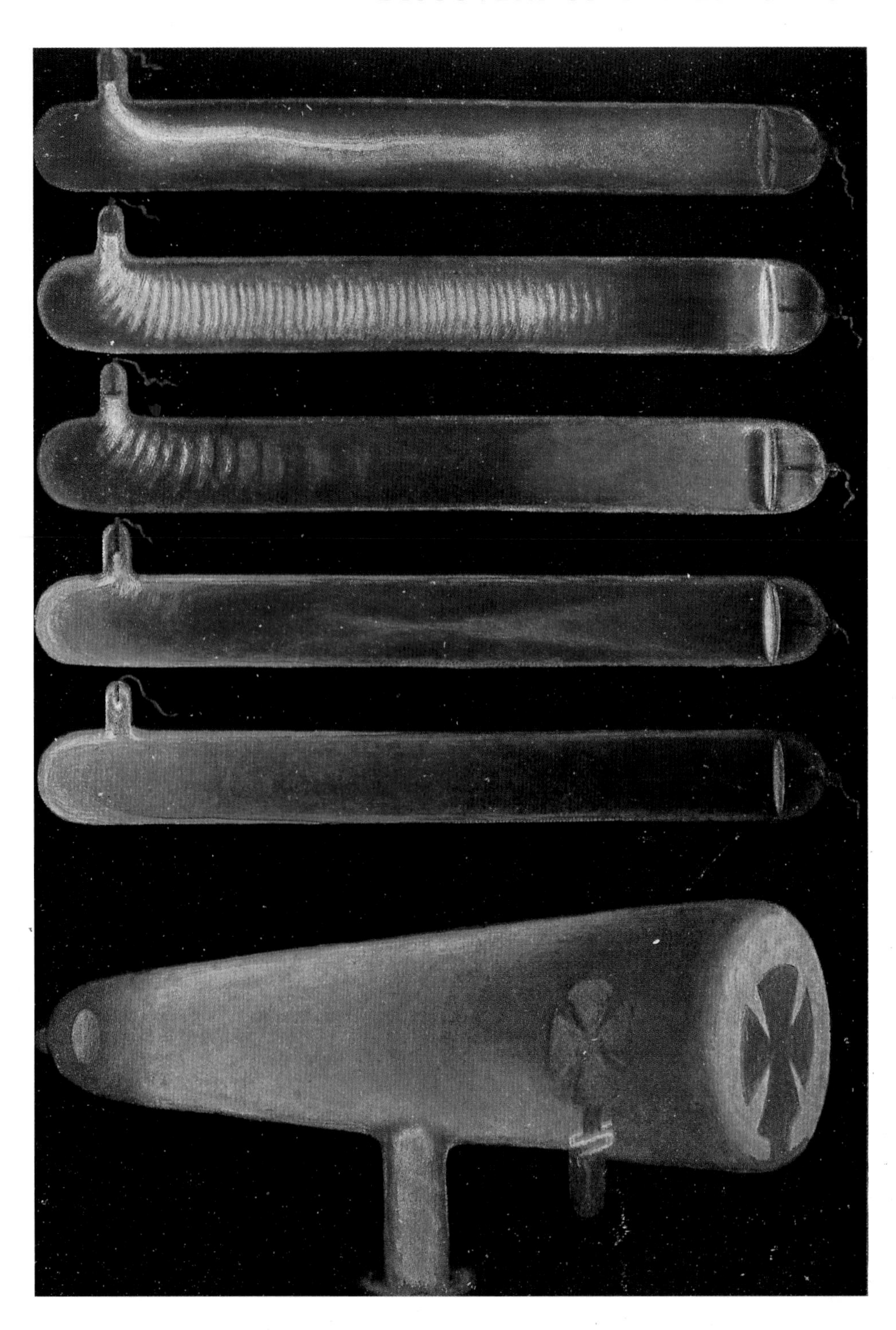

Cathode rays *The tubes placed in different electric fields form a variety of patterns. The principles discovered by J.J. Thomson 100 years ago govern the cathode-ray tube that is found in television sets today. In a television, electric and magnetic fields control the movement of a beam of electrons – or cathode ray. The beam makes dots on the fluorescent television screen and so builds up the picture.*

NEW MODELS OF THE ATOM

In 1882, aged 26, Joseph John Thomson, the son of a bookseller, won a popular science prize for an essay on the motion of vortex rings – such as smoke rings. His great discovery, however, was of the electron, and this overthrew the 2,400-year-old idea that atoms (if they existed at all) were hard and indivisible. In 1899 Thomson proposed a new model of the atom, picturing it as a sphere of uniform positive charge with the negative electrons embedded in it.

Philipp Lenard in France claimed in 1903 that the atom was almost entirely empty space. This was a radically new idea, but was soon to be confirmed. His atom had structures called "quants" – centres of negative electric force – which were closely coupled to positive charges, forming neutral pairs called "dynamides".

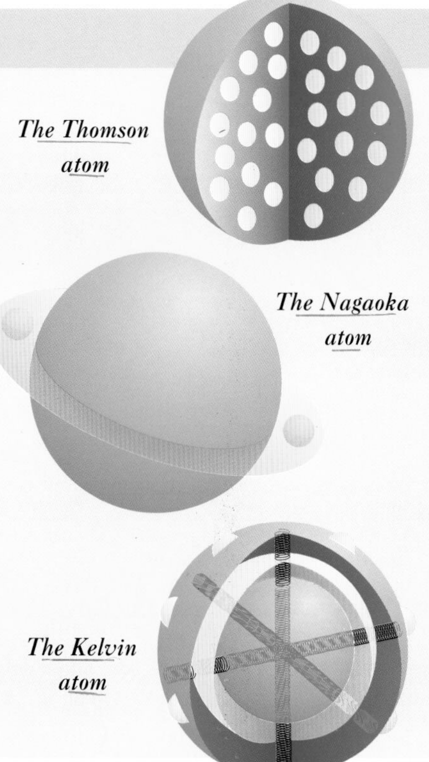

The Thomson atom

The Nagaoka atom

The Kelvin atom

In Japan in 1904, Hatari Nagaoka proposed a large positive charge concentrated in the centre of the atom, with the negative electrons orbiting around it in rings like those of Saturn.

In Scotland, William Thomson (no relation of J.J.), later to become Lord Kelvin, was a prolific atomic modeller, having first suggested that atoms were vortices in a frictionless fluid. Later (in 1905) he proposed atomic arrangements of concentric spheres linked by systems of "springs".

In Germany, a number of influential scientists – such as Ernst Mach, Heinrich Hertz and Wilhelm Ostwald – stubbornly resisted atomism. Atoms were not real, maintained Hertz, they were merely imaginary objects that were a useful means of explaining certain phenomena.

Sir William Crookes *in a caricature by "Spy", published in* Vanity Fair, *1903. He is shown holding the discharge tube that carries his name. The Crookes tube produces electron discharge patterns like those shown opposite.*

ELECTRON THEORIES

In 1892 the Dutch scientist Hendrik Anton Lorentz put forward the idea that matter contains many extremely small charged particles, the electrons. He came to this conclusion while he was looking at the relationship between Maxwell's electromagnetic radiation and the matter from which it comes. Lorentz claimed that light and other radiation is produced by vibrating electric charges.

Lorentz pictured electrons as hard spheres smeared with electric charge. His theory explained many electromagnetic properties of matter, but he did not ask whether his electrons had anything to do with the structure of matter.

Joseph John (J.J.) Thomson succeeded Maxwell as Professor at the Cavendish Laboratory in Cambridge, England. In 1897 Thomson finally settled the cathode ray controversy by showing that the rays were deflected by an electric field as well as by a magnet. Through his experiments he proved that the rays were particles. He also proved that they were negatively charged and about 2,000 times smaller than the smallest atom – that of hydrogen.

At first Thomson referred to his particles as "corpuscles", but later adopted the name electron. His historic conclusion was: "Thus we have in the cathode rays matter in a new state, a state in which the subdivision of matter is carried very much further than in the ordinary gaseous state".

Thomson's announcement of the discovery of the electron to the British Royal Institution on 30 April 1897 can be regarded as the birth of particle physics. Radioactivity and X-rays had already been discovered, but their connections with atomic structure were not yet understood. The electron was the first subatomic particle to be identified.

The electric charge of the electron was measured by the American Robert Millikan in 1911 using an elegant experiment. Millikan sprayed small drops of oil between two electrically charged horizontal plates, and then watched the drops through a microscope as the voltage on the plates was changed. He was able to measure the amount of electric charge they had taken up. This charge was always in multiples of one number, which was the charge on a single electron.

In 1900 a revolutionary idea in physics was born. Studying heat radiation, Max Planck concluded that radiant energy is not emitted in a continuous stream, but in small packets called quanta. Albert Einstein extended this idea, saying that light behaves like a burst of particles called photons.

Is light schizophrenic?

THE QUANTUM REVOLUTION

Towards the end of the nineteenth century, physicists were puzzled by how hot objects appeared to radiate heat energy. Experiments showed that the radiated heat depends solely on the temperature, not on the material the object is made of. Attempts to relate energy to temperature gave an absurd prediction. A so-called "black body" – a "perfect radiator" of energy – would give off an infinite amount of energy at the high frequency (ultraviolet) end of the electromagnetic spectrum.

This "ultra-violet catastrophe" remained a problem until Max Planck, Professor of Physics at Berlin, found a radical solution. On 14 December 1900 he reported to the German Physical Society that radiant energy is not continuously and infinitely divisible. Instead, the radiation emerges in bursts as small packets of energy he called "quanta" (from the Latin "how much"). The higher the frequency of the radiation, the more energy each quantum carries.

Ludwig Boltzmann, inventor of statistical mechanics and a giant of nineteenth century physics, had already (in 1872) used the idea of energy quanta, but not the word. He considered it to be a mathematical trick, but the approach probably influenced Planck.

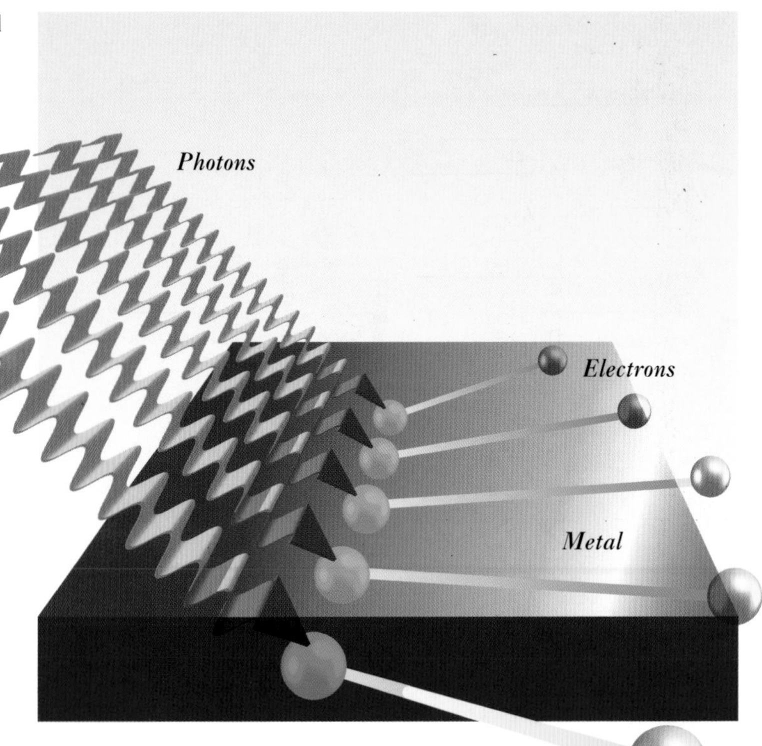

MAX PLANCK – AN ETERNAL OPTIMIST

"His work has given one of the most powerful of all impulses to the progress of science. His ideas will be effective as long as physical science lasts," Einstein wrote about Max Planck (1858–1947). Einstein regarded Planck as an almost saint-like figure.

Despite his scientific success, Planck's personal life was tragic. His oldest son was killed at Verdun in World War I, and both his daughters died soon after marriage. A younger son, Erwin, took part in the attempted 1944 coup against Hitler and was executed. Such events inevitably left their mark, but Planck's motto remained "*man muss Optimist sein*" (one must be an optimist). In 1918 he was awarded the Nobel Prize for Physics.

ACT OF DESPAIR

Planck called the quantum hypothesis an act of despair to explain experimental results. A physicist of the old school, he felt uncomfortable with quanta, and spent the next ten years trying to get rid of them, so that the theory could be reconciled with classical physics. "Many of my colleagues saw in this something of a tragedy", he later wrote.

Planck did, however, fully understand the implications of his discovery. "It might be the biggest thing in science since Newton", he confided to his son. The implications became evident when Albert Einstein, a 26-year-old failed schoolteacher who was then working at the Patent Office in Bern, took Planck's ideas one step further.

Photoelectric effects (left and below) *When light is shone onto photosensitive metals, electrons are knocked out of the metal atoms. If light was entirely wave-like in character, we would expect that turning down the brightness of the light would cause slower electrons to be emitted. In fact, even a very faint light produces fast electrons. This behaviour can only be understood if light is made up of particles. The effect can be imagined in terms of multiple jets of water aimed at a box full of ping-pong balls. When only one jet is in action, one or two balls still fly out at high speed.*

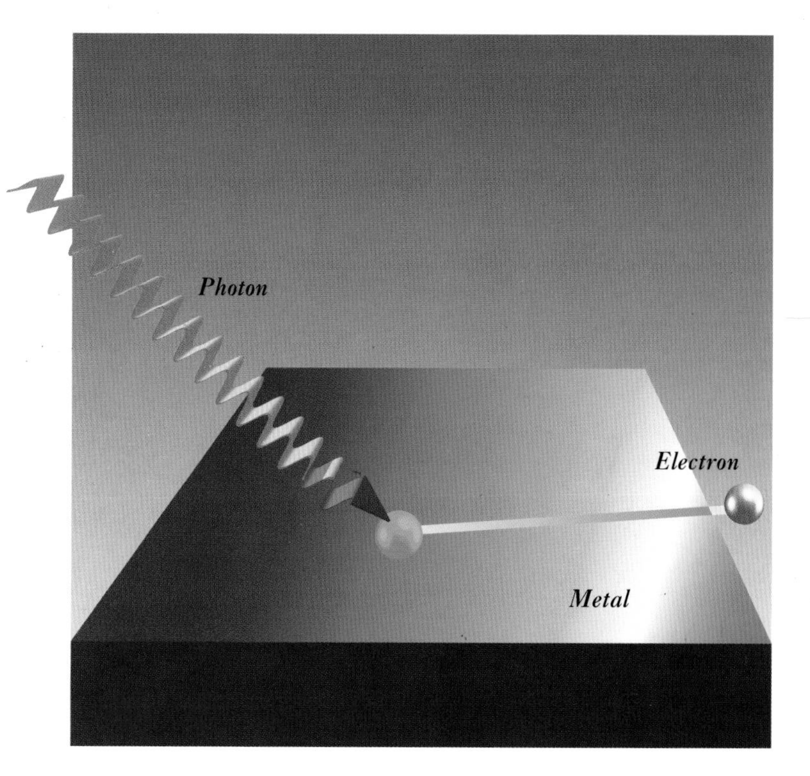

THE PHOTOELECTRIC EFFECT

The photoelectric effect had been known since the 1880s. Light occasionally knocked electrons off the surface of certain metals to produce a tiny electric current. In 1902, Philipp Lenard had shown that increasing the intensity of the light did not make the electrons fly off with greater energy, but produced a larger number of electrons with the same energy.

This was incomprehensible if light was made up of waves, as almost all physicists then believed. Einstein came to the rescue, showing that the effect could be easily understood if light behaved like a staccato of particles, like bullets from a machine gun. He later called the light particles "photons". The energy of a photon depends on the frequency (the number of oscillations per second) of the radiation; the higher the frequency, the more energetic the photon. The ratio between energy and frequency is a fixed number known as Planck's constant.

THE DUAL NATURE OF LIGHT

Einstein's new interpretation went one step beyond Planck, who had maintained that radiant energy was only emitted and absorbed in quanta, not that the radiation itself was discontinuous. Plank supposed that the quanta somehow joined together to produce waves, which then fragmented into quanta when absorbed.

Einstein knew he had created a dilemma. A century of careful experiments had apparently established that light consisted of waves, and these conclusions could not be ignored. For the next 20 years, Einstein struggled to understand what he called the "schizophrenic" character of light. How could something be particles and waves at the same time? It was like saying that a stone is the same as the ripples it makes on a pond. Could the photons, he speculated, be accompanied by "ghost" waves?

EINSTEIN'S TRIPLE

Albert Einstein, at the age of 26, published three landmark scientific papers, all appearing in the same volume of the German scientific journal *Annalen der Physik* in 1905. The first presented his explanation of the photoelectric effect, proving that light consists of particles (photons). The second paper explained Brownian motion, the random microscopic movement of particles suspended in water first observed by the Scottish botanist Robert Brown in 1827. Einstein derived an equation showing that Brownian motion was due to molecules of water crashing at random into the suspended particles.

The third paper described Einstein's special theory of relativity, which overturned common-sense ideas of time and space, and re-examined the notion of simultaneous events. Possibly because the ideas of special relativity, which explain what happens when velocities approach those of light, were initially difficult to appraise, it was the first paper on the photoelectric effect and not the third that won Einstein the 1921 Nobel Prize for Physics. However, Einstein's new view of space and time changed for ever our understanding of the Universe and its constituents on both small and large scales.

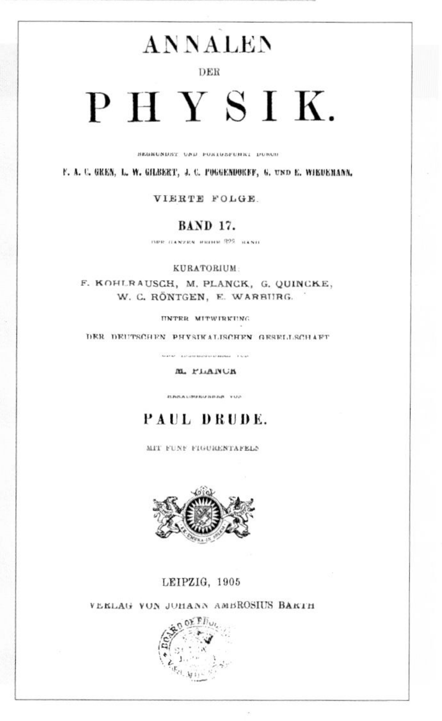
ANNALEN
DER
PHYSIK.

F. A. C. GREN, L. W. GILBERT, J. C. POGGENDORFF, G. UND E. WIEDEMANN.

VIERTE FOLGE.

BAND 17.

KURATORIUM:
F. KOHLRAUSCH, M. PLANCK, G. QUINCKE,
W. C. RÖNTGEN, E. WARBURG.

UNTER MITWIRKUNG
DER DEUTSCHEN PHYSIKALISCHEN GESELLSCHAFT

M. PLANCK

PAUL DRUDE.

MIT FÜNF FIGURENTAFELN

LEIPZIG, 1905
VERLAG VON JOHANN AMBROSIUS BARTH

Radioactivity became a powerful new tool in the hands of Ernest Rutherford. In 1911 he fired radioactive particles at a thin gold foil and found that the atom was mostly empty space, resembling a miniature solar system, with most of the mass concentrated in a tiny central nucleus.

The nuclear atom

THE DISCOVERY OF THE ATOMIC NUCLEUS

ALPHA, BETA AND GAMMA

In 1900, Paul Villard, a French chemist with parents of English origin, found a third kind of radioactivity that Rutherford had missed because it passed unnoticed through his apparatus. With alpha and beta radiation known, Villard logically called it "gamma rays". Alpha (α) and beta (β) rays are, in fact, particles; alphas are helium nuclei, betas are electrons. But gamma rays (γ) are true electromagnetic radiation. They are the most energetic form of electromagnetic radiation, with shorter wavelengths than X-rays. Gamma rays are often emitted at the same time as alphas and betas, allowing the radioactive nucleus to "settle down" into a less active state.

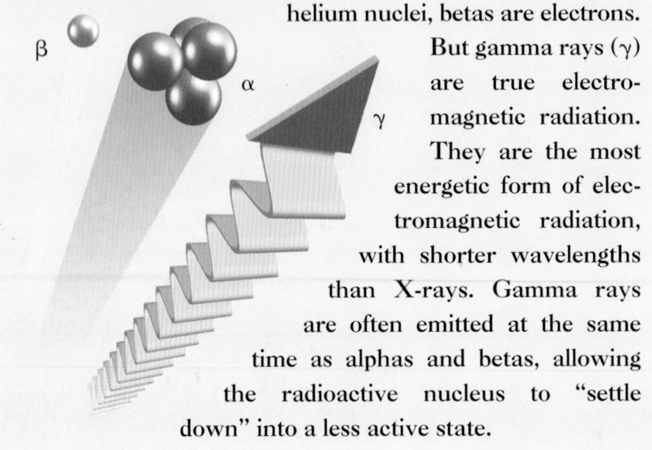

Ernest Rutherford came to England from New Zealand in 1895, at the age of 24. Back home he had built a pioneering radio-wave detector and during his first months at the Cavendish Laboratory, where he worked under Professor J.J. Thomson, he made remarkable progress on the study of radio waves. He was technically ahead of Marconi, the inventor of wireless telegraphy, and for a time held the world record for transmission and reception of radio waves – more than 3 kilometres (2 miles).

However, the news of X-rays and radioactivity made Rutherford give up his promising radio detector. Encouraged by Thomson, he threw his enormous talent and energy into this new field and soon made an important contribution. In 1898 he showed that uranium gave off not one but two kinds of radiation, which he called alpha and beta rays. They were not ordinary rays, but streams of high-speed particles, which could be distinguished from each other by their different penetrating powers.

The nuclei of some atoms are stable and live forever. Others, such as those of uranium, are unhappy and try to find a more stable composition by breaking down. This instability, with fragments flying off nuclei, is known as radioactive decay. Until 1932, these natural nuclear fragments were the only means of triggering new nuclear reactions.

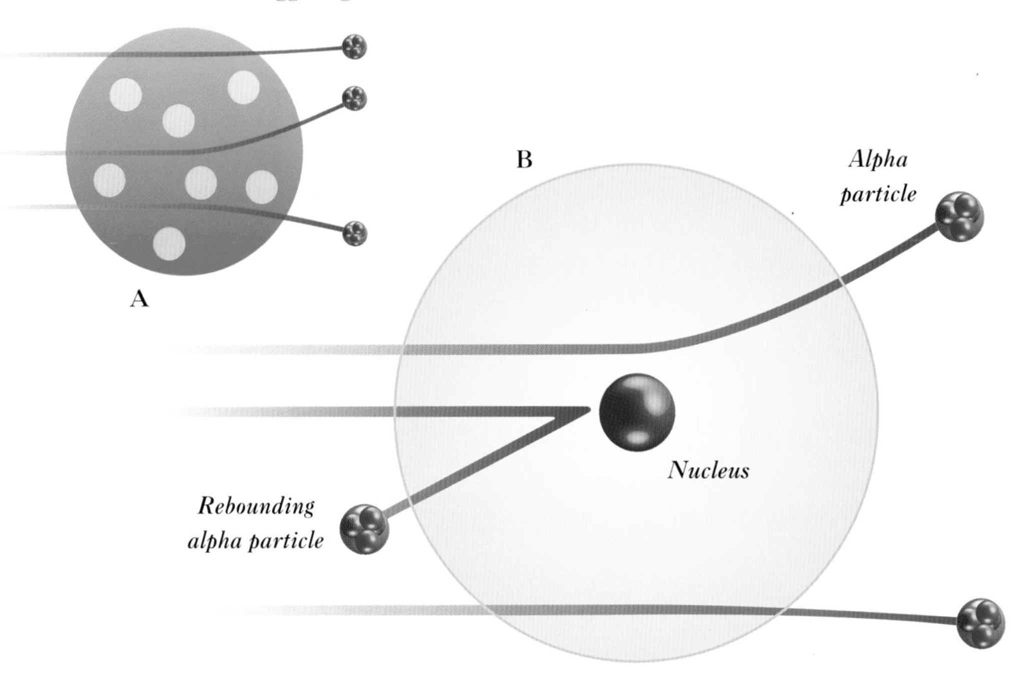

Strong recoil *Physicists at first visualized the atom as a sphere of positive charge with electrons embedded within it. In 1911, Rutherford fired alpha particles at gold atoms. But instead of passing straight through with only slight deflections as expected (A), some of the alphas bounced backwards (B). The rebounding alpha particles showed that more than 99 per cent of the mass of the atom is concentrated in a tiny central positively charged nucleus. Using basic physics ideas, Rutherford worked out exactly how the deflection of each particle depended on the distance of closest approach between the incoming particle and the nucleus. This "Rutherford scattering" was later refined, but remains essential for every student of modern physics.*

In 1898 Rutherford moved to Montreal, Canada, to become professor at McGill University. There he continued his research into radioactivity and in 1903, with chemist Frederick Soddy, studied how radioactive elements decay.

While beta radiation was quickly identified as streams of electrons, alpha particles were more of a mystery. Rutherford had a special affection for alpha particles and called them his "happy little chaps". They were for a time regarded as secondary X-rays, but in 1909 Rutherford managed to trap alpha particles inside a thin glass bulb and showed that they were actually helium nuclei.

NAVAL BOMBARDMENT

In a landmark 1911 experiment, Rutherford told his students to fire alpha particles from a radioactive source at a thin gold foil to see how they were deflected. The particles emerging from the foil were then patiently recorded, seen through a microscope as flashes on a fluorescent screen.

Most of the alpha particles went straight through the foil, but a few were strongly deflected and some even ricocheted off backwards as if they had slammed into something heavy. "It was as though a naval gun were fired at a piece of paper and the shell came right back to hit you", said Rutherford. After thinking about this strange result for several weeks, he came up with an explanation that changed our view of the microscopic world.

If most of the particles passed right through the foil, he reasoned, the atom must be made up mostly of empty space. But deep inside the atoms was a tiny central nucleus where most of the mass was concentrated, so that on the rare occasions when an alpha particle collided with a nucleus, it swerved violently. The electrons orbited around the nucleus, like planets around the Sun, so that the whole atom, concluded Rutherford, was some 10,000 times bigger than the nucleus. With this, the search for the extremely small took a major step forwards.

Ernest Rutherford *"All science is either physics or stamp collecting", Rutherford once remarked. A down-to-earth New Zealander, he was physically tall and strong with a loud voice to match. When he walked around singing "Onward, Christian soldiers" off-key, his staff knew things were going well. Few have had a greater impact on science in the 20th century. He dominated nuclear physics until his death in 1937, although he never fully grasped the potential of his nuclear ideas, dismissing the possibility of atomic energy as "moonshine". Rutherford's booming voice reflected great self-confidence. As his student Patrick Blackett, who was to win a Nobel Prize, later wrote, Rutherford "had not very much to be modest about".*

Quantum jump

BOHR'S MODEL OF THE ATOM

Although providing new insight, Rutherford's revolutionary nuclear atom idea was still based on the classical physics of Newton and Maxwell. However, it quickly became clear that atoms obey different rules. Niels Bohr produced a new model to account for how atoms behave.

Rutherford's solar-system model of the atom had a serious flaw. According to the laws of electromagnetism, rotating electrons should continually give off radiation and lose energy, and so would quickly collapse into the atomic nucleus. When, then, were atoms stable?

Either there was something wrong with the Rutherford model – or with convetional physics. A young Danish physicist called Niels Bohr chose the latter. The atomic world had different rules, he surmized, and in 1913 used Planck's quantum theory to develop an improved picture of the nuclear atom.

According to Bohr the atomic electrons could only occupy well-defined orbits corresponding to fixed energy levels. In these orbits the electrons circulate happily without losing energy. However, when an atomic electron received a kick-like impulse, it jumped to a higher energy orbit. Subsequently it fell back to its "ground state", sending out a light quantum (a photon) corresponding to the energy difference between the two orbits. Depending on the orbits involved, these photons could be visible or ultraviolet light, or X-rays.

This was the famous "quantum jump". The strange orbital shift of the electrons caught the public's imagination. It was as if a car struck by lightning in a city street suddenly disappeared into thin air and turned up on an expressway; after the storm, the car reverted to its original route, giving off a flash of lightning. Each atom had its own characteristic pattern of quantum jumps.

As well as showing why atoms were stable, Bohr's model of the atom also explained spectral lines. When light from a hot gas is shone through a prism, it comes out as a well-defined sequence of coloured bands, or spectral lines, that form a characteristic "signature" of the gas.

NIELS BOHR – SOCCER AND WESTERNS

Niels Bohr (1885–1962) played soccer in his youth and was a reserve goalkeeper for a leading Danish club. However, his reluctance to come out and meet attacks prevented him from taking a place in the first team. His brother Harald played in the halfback position and was a star in the Danish Olympic team that won the silver medal in England in 1908.

Despite his scientific brilliance, Bohr was a slow thinker. George Gamov related that on visits to the cinema Bohr would keep asking stupid questions. Although addicted to American Westerns, he frequently had problems in following the plot. His accumulated experience from these Westerns led him to develop a "theory" of defensive shooting – that a voluntary decision is always slower than an instinctive response. He confirmed his hypothesis with the help of the Dutch physicist Hendrik Casimir and some toy pistols.

Atomic stepladder (below and right)
An incoming light quantum (below), a photon, is absorbed by one of the captive electrons in an atom, pushing the electron into an orbit with higher energy further from the nucleus. Unhappy with its acquired energy, the excited atom then reverts to its original configuration, spitting out the photon (right) as the electron returns to its original orbit closer to the nucleus. For each type of atom, the possible energies of the electrons are fixed, like rungs on a ladder. As the electrons shuffle up and down this energy ladder, they absorb or emit photons. This produces a characteristic pattern of spectral lines – an atomic "fingerprint" – each line corresponding to a particular energy jump by an orbital electron.

Bohr explained these bands by atomic electrons shuffling between their different possible orbits. Even hydrogen, the simplest atom with just a single electron, nevertheless has a complicated spectrum. Bohr showed that this pattern corresponded exactly to the sequences of possible orbital jumps that a single electron can make.

Arranging electrons in rings or "shells", like the layers of an onion, Bohr said each shell could not accommodate more than a certain number, and the lower-energy shells were filled first, starting with those nearest the nucleus.

ATOMIC SIGNATURE

In Bohr's picture, elements with the same number of outer electrons had similar chemical properties. Now the strange pattern of Mendeleev's periodic table could be understood in terms of atomic electron arrangements.

In Bohr's picture of the atom, electrons sit happily in fixed orbits until some external "kick" pushes one out to another orbit, further from the nucleus. Eventually the electron shifts back to its original orbit, emitting a flash of light. The bands of colour – "spectral lines" – seen when light from a hot gas is passed through a prism are due to these electron jumps. Each element has its characteristic spectral signature, so that the analysis of starlight, for example, shows what the star is made of.

But the explanation of Bohr's electron assignments only became clear when Wolfgang Pauli in 1926 discovered the "exclusion principle", which governs how particles like electrons coexist. Pauli saw that no more than one electron can be put into each possible quantum slot. By then a new generation of physicists, spurred by Bohr's success, was probing the deep implications of quantum theory and staged a second revolution, "quantum mechanics".

Bohr was also an avid traveller, visiting Rutherford in Manchester for the first time in 1912. Personal contacts were extremely important in understanding and exploiting the new quantum ideas, and Bohr's Copenhagen school became a focus for European and world science. He was awarded the Nobel Prize for Physics in 1922. Unlike many of his colleagues, he continued to be scientifically productive for a long time, producing new nuclear-physics ideas when he was well into his 40s. After World War II, he played a major role in reshaping European physics in a war-torn continent.

A fuzzy world of uncertainty

THE QUANTUM MECHANICAL PICTURE

In the 1920s a new theory, called quantum mechanics, gave a better understanding of the atom. But it also made the atom look more fuzzy. Instead of being fixed in orbits, electrons were smeared out. Uncertainty and statistical probabilities were introduced as basic properties of nature.

Physics in the early 1920s was in poor shape. Bohr's atomic model worked fine for the hydrogen atom with its single electron, but was difficult to extend to more complex atoms. Also Bohr could not explain how the electrons jumped from one orbit to another, and was the first to admit that his atomic model was far from satisfactory.

One talented young German, Werner Heisenberg, openly called the Bohr atom "a peculiar mixture of mumbo-jumbo and empirical success". Not at all upset by this and aware of the shortcomings of his theory, Bohr invited Heisenberg to visit Copenhagen and see if he could develop a more consistent and useful picture of the atom. It was on a journey home from Copenhagen in the summer of 1925, during a brief stay on the island of Heligoland to recover from hay fever, that Heisenberg had a remarkable idea.

The Bohr atom failed, said Heisenberg, because it was based on things that cannot actually be observed – for example, electron orbits. Instead he took an abstract approach, and described the atom in a new, purely mathematical way, called "matrix mechanics", with the electrons permitted only to have energies in a series of definite values.

WAVES OF PROBABILITY

At roughly the same time, as so often in physics, parallel investigations were taking place in another part of the world. For Christmas 1925, the Austrian scientist Erwin Schrödinger went skiing with a girlfriend, leaving his wife in Zurich. However, he never did much skiing because up in the Swiss Alps he "was distracted by a few calculations".

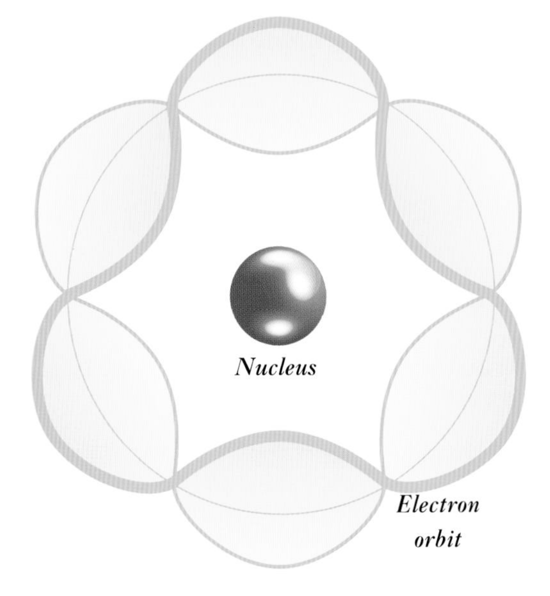

A new idea was puzzling Schrödinger. The Frenchman Louis de Broglie had suggested in 1923 that because light waves sometimes show particle properties, so matter particles, such as electrons, should sometimes behave like waves. Schrödinger was determined to put de Broglie's idea into a mathematical form. The result was a new method of describing the atom, called "wave mechanics".

In Schrödinger's wave-mechanical atom, the electrons were confined in wave patterns around the nucleus. Only complete wave patterns could be accommodated. This explained the mysterious quantum jumping, which Schrödinger had detested. "Surely you realize," he had once said to Bohr, "that the whole idea of quantum jumps is bound to end in nonsense."

PHYSICS FOR BOYS

The radical ideas of the new quantum theory appealed to fresh, uncluttered minds. Werner Heisenberg (pictured left), born in Duisberg, Germany, in 1901, was just 26 when he became famous as the result of his uncertainty principle. Most of the other contributors, such as Paul Dirac, Wolfgang Pauli and Pascual Jordan, were in their early twenties. As a result the theory was contemptuously called "boy physics". A notable exception was Erwin Schrödinger, at 37 already a respected figure. However, the mature Schrödinger had difficulty in interpreting his own invention, wave mechanics.

QUANTUM MECHANICS IS BORN

Schrödinger's simple atomic picture was put into question, in turn, when Max Born interpreted the electron waves as probability distributions. Quantum mechanics, said Born, can only calculate the probabilities of finding electrons and cannot give their exact location. It is better to think of electron "clouds" of varying thickness rather than fixed orbits.

In 1927, struggling with the mathematical description of an electron's path through a cloud chamber (see page 43), Werner Heisenberg discovered a dramatic new quantum effect, now called the Heisenberg uncertainty principle. This says that it is impossible to know simultaneously both the position and the momentum (velocity and mass) of a particle. The world at the level of atoms is by its very nature full of fuzziness. The uncertainty principle also says that "empty" space – the vacuum – is filled with quantum clutter constantly flashing on and off. These tiny bursts of transient energy transmit messages between particles and play a fundamental physics role as force carriers.

These diverse approaches to atomic physics were difficult to reconcile, but the Solvay physics meeting in Brussels in 1927 sought to bring some accord. "We all stayed at the same hotel", recalled Heisenberg, "and the fiercest arguments took place not in the conference but during the hotel meals." The result was a surprise to all: Heisenberg's matrix mechanics, Schrödinger's wave mechanics, Born's probability distributions, and the uncertainty principle were, in essence, all describing the same thing. This new view of the atom came to be known as "quantum mechanics".

Einstein was never comfortable with the idea that quantum events, ruled by probability, had an element of randomness. He tried continually to come up with objections to the uncertainty principle. "God does not play dice!" Einstein often remarked.

The Schrödinger atom (above left) *Electrons, although normally particles, also behave as waves. Schrödinger believed that when an electron orbits a nucleus, its wave is confined in the orbit, producing a "standing wave" – something like the patterns seen on a skipping rope – vibrating up and down, but without moving along.*

The fuzziness of nature (above) *The Heisenberg uncertainty principle states that it is impossible to know exactly at the same instant both where a subatomic particle is and how fast it is moving. This is something like trying to take a picture of a moving athlete (such as this one winning an Oxford/Cambridge versus Harvard/Yale race in 1923). If we try to see his* motion, *then his* position *in space becomes blurred.*

MATTER WAVES

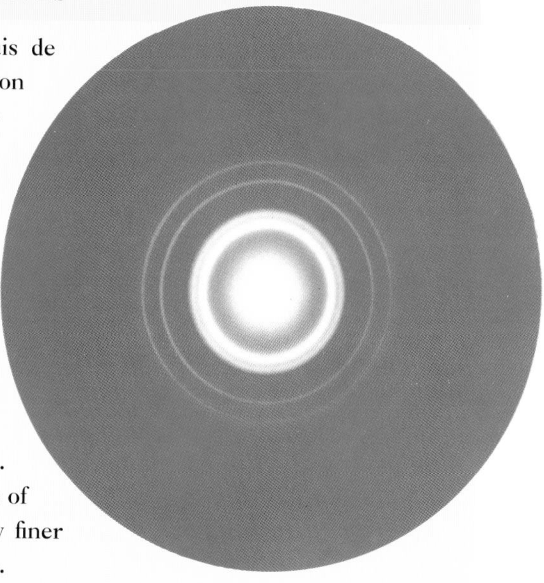

The matter waves predicted by Louis de Broglie were seen in 1927 when Clinton Davisson and Lester Germer in the United States and George Thomson (the son of J.J.) in Britain showed that electrons could be diffracted by crystals and metal foil (as shown here), in the same way that light is diffracted by a fine mesh. By 1931 a German, Ernst Ruske, had built a new kind of microscope using a beam of electrons in place of light to illuminate the object being observed. Electrons have wavelengths thousands of times shorter than light and can show finer details, magnifying up to a million times.

Smashing the atom

UNRAVELLING THE NUCLEUS

Since the time of the ancient Greeks, atoms had been thought to be the ultimate indivisible constituents of matter. Early in the twentieth century physicists discovered that this was not the case – inside the atom was a nucleus – and then that the nucleus itself was divisible.

In 1914 at Manchester University in England, Rutherford's assistant Ernest Marsden was continuing the scattering experiments that had led to the discovery of the nuclear atom in 1911. He noticed that fast-moving hydrogen nuclei, which he called "H particles", were sometimes seen when alpha particles were in action. He supposed the H particles came from the radioactive alpha-particle source.

Joining the hunt for the H particles, Rutherford at first thought they were an unknown light gas. The great man's research was held up by World War I, but he was still able to spend some time in the laboratory. Absent from one submarine-committee meeting, he apologized with the legendary remark: "If, as I have reason to believe, I have disintegrated the nucleus of the atom, this is of greater significance than the war."

MODERN ALCHEMY

In 1919, with the war over, Rutherford resumed the investigation. He fired alpha particles into a container filled with nitrogen gas, and a fluorescent screen behind the apparatus monitored what happened. Normally the alphas were absorbed by the nitrogen, but occasional flashes showed that new particles, more penetrating than alphas, came flying out. Rutherford concluded that nitrogen nuclei struck by alphas had changed into oxygen, releasing at the same time a hydrogen nucleus. Speculating that these hydrogen nuclei were the building-blocks of all nuclei, Rutherford called them "protons" (meaning "first particles").

The experiment was the first time anyone had seen the transformation of an element; the old alchemists' dream had come true. Over the next few years physicists kept firing alpha particles at nuclei, triggering more nuclear reactions. But before new techniques arrived, all they could do was count flashes on a fluorescent screen viewed through a microscope.

It was not until 1925 that Patrick Blackett, using a cloud chamber (see page 43) of his own design in Rutherford's laboratory in Cambridge, made the first visual record of a nuclear disintegration. The photograph he produced showed the tracks of particles involved in Rutherford's transformation of a nitrogen nucleus to an oxygen nucleus. In 1948 Blackett received a Nobel Prize for his work.

However, natural alpha particles were inefficient nuclear "bullets", and Rutherford called for artificial sources, providing greater energies than natural radioactivity. John Cockcroft, joining Rutherford's Cambridge laboratory in 1924, set out to use protons accelerated by a high-voltage electric field. Although Rutherford realized that higher-energy particles were necessary, he was not convinced that this needed expensive machines. This was one of his few scientific mistakes, and the baton passed to the United States (see page 50).

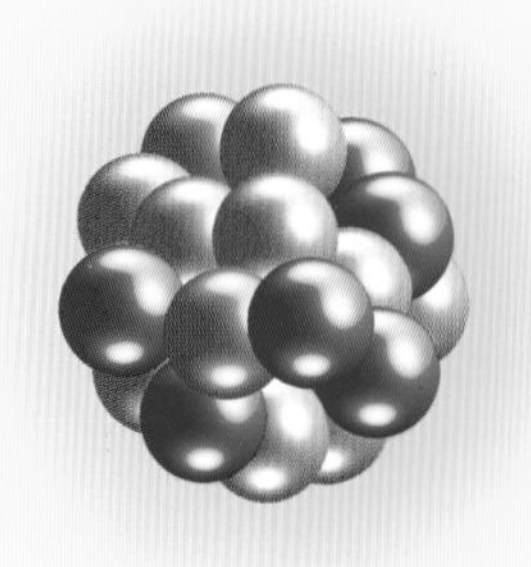

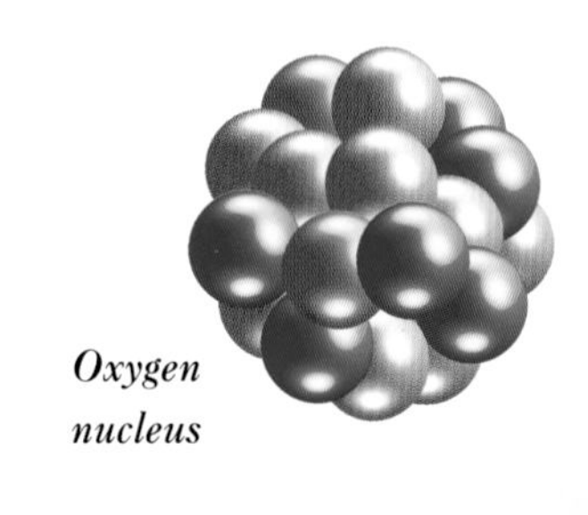

Rutherford's nuclear transformation
An incoming alpha particle hits a nitrogen nucleus (seven protons and seven neutrons), which transforms into an oxygen nucleus (eight protons and nine neutrons). The interaction liberates a proton, which zooms off downwards to strike a fluorescent screen, creating a visible flash, while the heavy oxygen nucleus is left behind.

"I KNOW AN ALPHA WHEN I SEE ONE"

In 1932, John Cockcroft and Ernest Walton were spending most of their time repairing vacuum leaks in the connections of their delicate glass apparatus. Eventually they coaxed it into operation and saw the first flashes on the fluorescent screen, showing that alpha particles were being produced. It was clearly time to call their boss. After a bit of trouble squeezing Rutherford's bulky frame into the tiny observation hut at the bottom of their apparatus, he began to bellow instructions: "Switch off the proton current! ... Increase the accelerating voltage! ... " After a while he emerged from the hut. "Those scintillations look like alphas," he said. "I should know an alpha-particle when I see one."

"WE'VE SPLIT THE ATOM!"

At first the energies required to break up a nucleus seemed too high. But George Gamov, a young scientist visiting Cambridge from Leningrad, claimed that lower-energy protons could drill into nuclei owing to an effect called quantum tunnelling. In this way Cockcroft and his collaborator Ernest Walton achieved the first completely artificial nuclear transformation in 1932.

In their experiment protons, ripped out of hydrogen gas and accelerated by the 800,000 volts of an ingenious high-tension system, were hurled down a vertical accelerator tube towards a lithium target. Sitting in a lead-shielded hut beneath the tube, Walton watched through the microscope as lithium nuclei changed into helium. The usually quiet Cockcroft reportedly rushed out of the laboratory into the street, shouting "We've split the atom! We've split the atom!" Newspapers speculated that scientists had now mastered unlimited atomic power that could "blow up the world".

Although the Cockcroft-Walton technique was soon overtaken in the race for ever higher particle energies, for 50 years their method was used to give electrically charged particles their initial energy kick before the newer, bigger machines took over.

The first particle accelerator ***Ernest Walton watching a fluorescent screen beneath the particle accelerator that he built with Cockcroft in 1932. Flashes on the screen indicated that alpha-particles were being produced as lithium was transformed into helium by high-energy protons.***

Hidden forces in the nucleus

PROTONS, NEUTRONS AND BINDING ENERGY

For a very long time the composition of the atomic nucleus was a puzzle. After a hunt that lasted 12 years, James Chadwick finally found the missing piece, the neutron. The discovery was to launch nuclear physics along the road that ultimately led to the development of the atomic bomb.

In 1920 Ernest Rutherford suggested that a third particle, in addition to the proton and the electron, existed inside the atom. At that time scientists thought that electrons not only circled the nucleus, but could also exist inside it. These internal electrons, they thought, were the source of the particles seen in beta radioactivity. Rutherford claimed that a proton and an electron sometimes could get together to form a neutral particle, which he called a neutron. Immediately he began to search for the particle, helped by his dour assistant, James Chadwick.

The initial searches for the neutron were unsuccessful, however, and the composition of the nucleus continued to baffle Rutherford throughout the 1920s. Meanwhile Chadwick, with a great amount of patience and staying power, continued his painstaking hunt for the elusive neutron.

PENETRATING RAYS

In 1930 the Germans Walther Bothe and Herbert Becker discovered that bombarding beryllium with alpha particles produced a penetrating radiation that passed through 10 centimetres (4 inches) of lead. They thought it was gamma radiation, until 1932 when Irène Curie (daughter of Marie) and her husband, Frederick Joliot, showed that the radiation knocked protons out of hydrogen atoms.

Repeating the Joliot-Curie experiment, James Chadwick came to another solution. The penetrating rays, he assumed, were particles. By using ordinary mechanics he deduced that the protons flying off with high speed had each been hit by a particle of about the same mass.

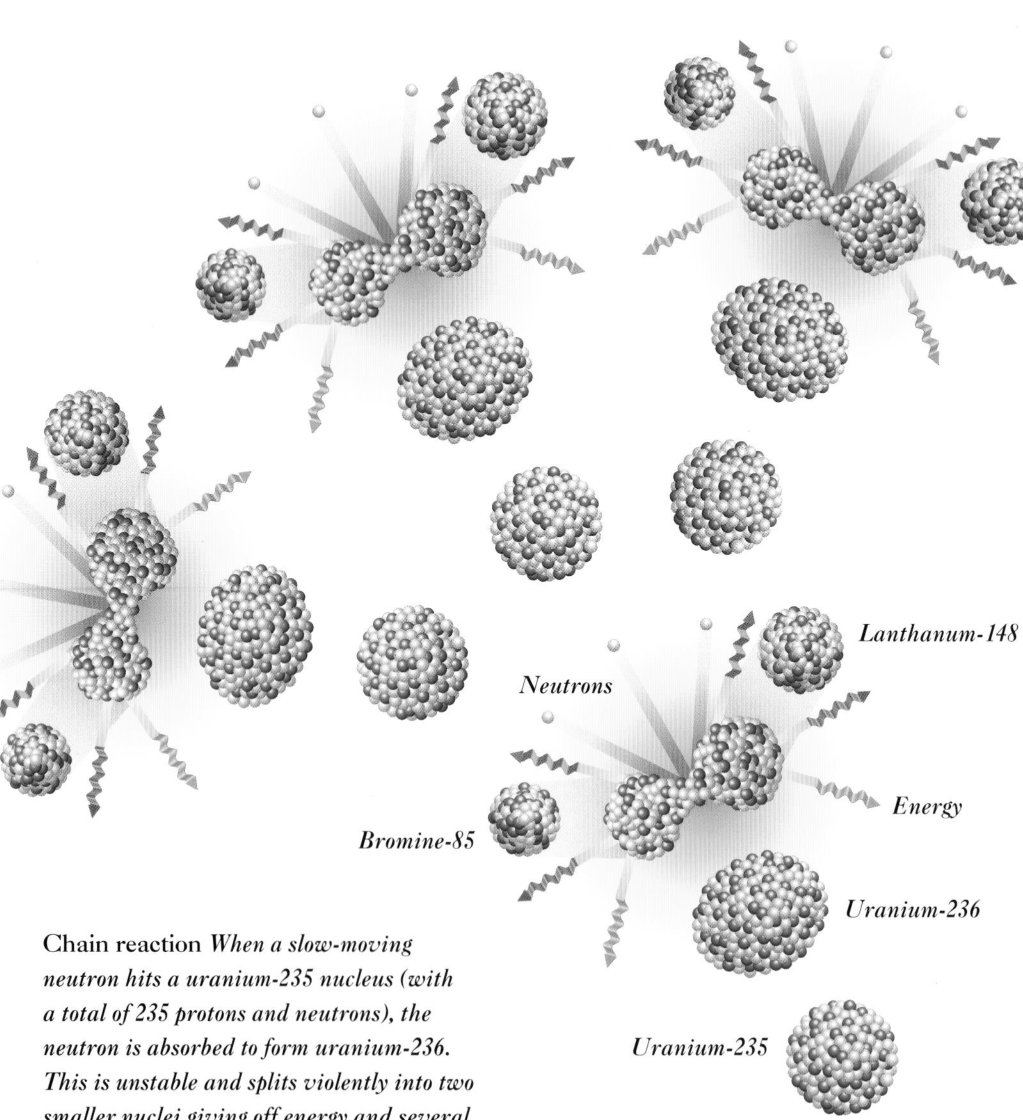

Chain reaction *When a slow-moving neutron hits a uranium-235 nucleus (with a total of 235 protons and neutrons), the neutron is absorbed to form uranium-236. This is unstable and splits violently into two smaller nuclei giving off energy and several neutrons. Each neutron released can, in turn, smash apart another uranium-235 nucleus. The reaction multiplies at lightning speed to give a chain reaction, releasing the huge force of a nuclear explosion in a fraction of a second. In a nuclear reactor, the explosion is controlled by ensuring that on average no more than one of the neutrons released hits another uranium-235 nucleus.*

The first nuclear bomb *The test of the world's first nuclear bomb (also known as the "atomic bomb"), pictured here, took place at Alamogordo Air Base in New Mexico, United States, on 16 July 1945. As early as 1907 Max Planck had suspected that huge amounts of energy might be locked up inside atomic nuclei, but his prediction went largely unnoticed.*

THE WORLD'S MOST FAMOUS EQUATION

In 1905, Albert Einstein published the equation $E=mc^2$, which can be read as energy (E) is equal to mass (m) multiplied by the speed of light (c) multiplied by itself (squared) – light travels at about 300 million metres (186,000 miles) per second. From the everyday point of view, energy and mass are very different things. Einstein showed that in fact they are equivalent and can be interchangeable. Mass turns into energy in a nuclear explosion.

THE NEUTRON

The particle discovered by Chadwick was the long-sought neutron, but it was a particle in its own right, not a combination of a proton and an electron. The nucleus, it was then realized, contains no electrons, only protons and neutrons. With no electric charge, neutrons are not affected by the electrical forces inside the atom and can easily pass through matter.

In 1934 Enrico Fermi, working in Italy, discovered that neutrons could be slowed down in substances such as water or paraffin. These slow neutrons were more easily captured by nuclei, giving more possibilities for artificial nuclear reactions that produced new radioactive variants (isotopes) of the original nuclei. These radioisotopes soon became useful tools in biology and medicine.

NUCLEAR FISSION

In Germany, Lise Meitner and Otto Hahn had been carefully analysing the products of radioactive decay. Meitner, a Jew, was forced to flee Germany in 1938. Hahn continued their work, now with Fritz Strassmann, and discovered that uranium bombarded by neutrons could split in two. Never before had radioactive decay been seen to produce such a drastic change in the composition of a nucleus.

Hahn wrote to Meitner, who relayed the news to her physicist nephew Otto Frisch. Meitner and Frisch called the new process "fission". As well as two fragments, fission also produced neutrons and energy. These neutrons could split more uranium nuclei, setting up a "chain reaction", the process that was the basis of the first type of nuclear bomb.

Something in the air

COSMIC RAYS

After the discovery of radioactivity, physicists found a new penetrating radiation, this time coming from outer space. They called it cosmic rays, and in the 1920s it set the stage for the next surprise.

Early twentieth-century physicists measured radiation from elements such as uranium with an instrument called the electroscope. The instrument consisted of two gold leaves that, when given an electric charge, stood apart. Radioactive emissions knock electrons off atoms in the air, making it slightly conducting, or "ionized". When an electroscope was placed in ionized air, the leaves slowly came together as the charge leaked away.

Curiously, air seemed to be slightly conductive even when there was no radioactive material around. The effect, reported by C.T.R. Wilson in 1900, was at first thought to come from radioactive substances in the Earth.

To test this, in 1910 Theodor Wulf, a Jesuit priest, took an electroscope to the top of the 300-metre (1,000-foot) Eiffel Tower. The decrease in the ionization was considerably less than expected and Wulf assumed that the radiation coming up from the ground was competing with other radiation coming down from above.

After reading about Wulf's work, Austrian physicist Victor Hess, with the help of the Austrian Aero Club, embarked on a series of daring balloon ascents in 1911. Using balloons filled with explosive hydrogen gas, he went up to a record height of 5,350 metres (17,550 feet). Sitting in an open wicker basket, shivering from the intense cold and breathless from lack of oxygen, he nevertheless managed to make a series of careful measurements.

He found that the ionization increased rapidly, and at 5,000 metres (16,500 feet) was several times more intense than at ground level. "From this I concluded that the ionization was due to a hitherto unknown radiation of extraordinarily high penetrating power which entered the atmosphere from space," he said in his lecture on receiving a Nobel Prize for this work in 1936.

Hess also tried to find out exactly where the extra-terrestrial radiation came from. On 12 April 1912 he made a balloon flight during a near-total solar eclipse of the Sun, but saw no decrease in activity.

THE CREATOR AT WORK

Hess's discovery aroused widespread interest, but World War I interrupted research. After the war studies resumed on both sides of the Atlantic, notably by the American Robert Millikan, who in 1925 coined the name cosmic rays, since the radiation appeared to come from the cosmos in general.

Millikan believed cosmic rays were a form of electromagnetic radiation more energetic than gamma rays, and was the "birth cry" of new matter being created at the edge of the Universe. "The creator", he claimed, "is still on the job."

Later it was found that cosmic rays are high-energy particles, mostly protons. They reach the upper atmosphere from different sources in our Milky Way galaxy, including the Sun, and possibly further afield. Colliding with atoms in the upper atmosphere, cosmic rays create a harmless shower of secondary particles which cascade to the ground. Although cosmic rays have been a valuable research tool to explore the world of elementary particles, their origin is a mystery (see page 114).

SUBATOMIC CLOUD TRAILS

In 1911 the Scottish physicist Charles Thomas Rees Wilson – C.T.R. for short – invented the "cloud chamber", an apparatus that revealed where subatomic particles had passed. Based on the principles of cloud formation, the chamber became an invaluable tool in nuclear and cosmic ray research.

Wilson had studied meteorology and knew that when air saturated with water vapour cools, the vapour condenses into droplets. This is how clouds form. The condensation is triggered by the slightest impurity or disturbance.

Wilson's cloud chamber consisted of a glass cylinder that contained a mixture of air and water vapour. The chamber had a piston which, when it was suddenly pulled out, caused the air in the chamber to expand and cool. When a particle passed through the chamber, water droplets from the supersaturated vapour condensed around the wake left in the particle's path, similar to the trail left by an aeroplane across the sky on a clear day. These microscopic clouds revealed the tracks of subatomic particles, such those of cosmic rays.

KEY

π^+	positive pion
π^-	negative pion
π^0	neutral pion
γ	gamma ray
e^-	electron
e^+	positron
N	nucleus of an atom
n	neutron
p	proton
ν	neutrino
μ^-	muon
μ^+	antimuon

Cosmic rain *The high-energy particles of cosmic rays crash into atomic nuclei in the upper layers of the atmosphere. These nuclear collisions produce showers of lower-energy but unstable secondary particles, many unknown at the time of the early measurements, which gradually decay as they descend. To intercept the higher-energy particles, physicists send their detectors to increasingly higher altitudes – tall buildings, mountains, airplanes, balloons and satellites. Other particles are best detected deep below Earth's surface.*

Mirror images of our world

ANTIMATTER AND POSITRONS

In 1932 all matter seemed to be built from three fundamental particles: protons, electrons and neutrons. But in that year a new particle, the positron, was discovered in cosmic rays. It was the first example of antimatter, a mirror-image of the ordinary matter of the everyday world.

Paul Dirac, a gifted English physicist who originally trained as an electrical engineer, developed a powerful equation that combined quantum mechanics with Einstein's special theory of relativity. As with many mathematical equations, Dirac's had more than one solution. One corresponded to the ordinary electron, but the other solution seemed to represent an electron with negative energy – whatever that might mean.

Werner Heisenberg and other quantum pioneers were worried by this negative energy. It did not correspond to anything in the physical world and yet seemed unavoidable because the mathematics of Dirac's equation was correct. In fact, Heisenberg initially called this "the saddest chapter of modern physics". But it was to end in intellectual triumph.

Dirac desperately wanted to find out what his negative-energy solutions meant. But in 1929, with his imagination run dry, he put the problem aside and went on long trips to America and Japan. On his return he had a solution – but a bizarre one. It was a theory of holes.

The idea was that electrons with negative energies are real – in fact there are electrons in all possible negative-energy states. We are surrounded by a "sea" of these electrons. They are normally unobservable, just as the air surrounds us, yet is normally unobservable. However, a vacancy or "hole" in this sea of electrons would occasionally occur. In an electromagnetic field, these holes would look like particles with positive charge. Dirac's idea was a dramatic example of the power of mathematics in a domain where human intuition is unreliable.

ANTIMATTER

In 1898 the British physicist Arthur Schuster made a remarkable anticipation of antimatter, surmising that there might be atoms with properties exactly opposite to those of ordinary atoms. Such "anti-atoms" would attract each other gravitationally, he thought, but might be repelled by ordinary matter. Collecting enough antimatter to do "simple" experiments is still a major challenge.

Moment of birth *Particle–antiparticle pairs are frequently created from high-energy radiation. This artificial picture simulates what happens when two electron–positron pairs are formed in quick succession from a burst of gamma rays (yellow). A magnetic field forces the paths of the electrons (green) and positrons (red) to curl in opposite directions. Particles and antiparticles can also annihilate each other when they collide, creating a burst of radiation.*

Antiworlds *It is conceivable that whole mirror-worlds of antimatter could exist. Matter and antimatter annihilate each other when they meet – much as the complementary colours in these portraits by Andy Warhol of Marilyn Monroe and her anti-Marilyn counterpart would neutralize each other if blended. So it would be vitally important to know whether a visitor from outer space came from a world of matter or antimatter before shaking its hand! The annihilation of matter and antimatter into energy now provides a valuable new window on the most basic physics of all, that of the vacuum, or void.*

THE PHILOSOPHERS' DREAM

Dirac jumped to a premature conclusion, identifying his electron holes with protons, the only positive particle that was known at the time. Heisenberg said the proton idea should not be accepted until a theory could be developed that would describe protons and electrons as two variants of the same particle. Dirac liked this unification idea, calling it "the dream of philosophers", but knew from the start that it would not succeed. The proton was more than 1,800 times heavier than an electron and was simply too big to fit into an electron hole.

Dirac abandoned the "rather sick" proton theory, and in 1931 concluded that the holes described particles with the same mass as the electron, but opposite electric charge. He initially called this new particle the "anti-electron".

Dirac was not particularly concerned with finding his anti-electron, but in 1932 the American Carl Anderson, unaware of Dirac's new ideas, found some strange particle tracks in cosmic rays. The tracks looked exactly like those of particles identical with electrons, except that they had opposite electric charge. They were too small to be protons.

Anderson tried all kinds of explanations. For example, though cosmic-ray particles normally come downwards, might they sometimes travel upwards, making the electron look as though it had the "wrong" charge? But slowly he was forced to the conclusion that the tracks revealed an unknown lightweight particle: a positive electron. Anderson called it the positron.

Learning of the positron discovery, Dirac went on to predict that there must also be an antiproton. This particle was discovered, though not until 1955. In fact, it has proved that in general every type of subatomic particle has its own antiparticle (although in some cases the particle is identical with its own antiparticle). Dirac's "donkey electrons", as critics called them because they insisted on going the "wrong" way in an electric field, were the harbingers of a whole new class of matter.

PAUL DIRAC – A TIGHT-LIPPED GENIUS

Paul Dirac, the father of antimatter, never talked much. He dreaded reporters and tried hard to avoid them when he travelled to Stockholm in 1933 to receive his Nobel Prize, which he shared with Erwin Schrödinger. Dirac's silence was attributed to the fact that his father, whose family was French-speaking, forced his children to speak French while they were at home. Not able to express himself well, Paul preferred to keep silent.

Dirac was imperturbable. During a US trip he was in someone's house when a fuel leak from from the heating system caught fire, threatening to burn down the building. Dirac calmly suggested they went outside and closed all doors behind them. The fire promptly went out for lack of oxygen.

While visiting Wisconsin University in 1934, Dirac was interviewed by an eager reporter, whose article described the encounter: "A mathematical physicist or something...who has been pushing Sir Isaac Newton and Albert Einstein off the front pages... A pleasant voice said 'Come in.'... This sentence was one of the longest of the interview... Dirac seems to have all the time in the world and his heaviest work is looking out of the window.

'Professor... will you give me the lowdown on your investigations?'

'No.'..."

Several more questions similarly drew a complete blank.

"Do you go to the movies?" asked the exasperated reporter.

"Yes."

"When?"

"In 1920." was Dirac's laconic reply.

Much ado about almost nothing

THE GHOSTLY NEUTRINO

The weirdest of all known particles is the neutrino, first predicted by Wolfgang Pauli in 1931. Close to being nothing, the neutrino has no electric charge, its mass – if it has any – is too small to measure, and it can pass unhindered across the entire breadth of the Universe.

For many years, beta-radioactivity was dogged by what physicists termed an "energy crisis". In beta decay, a nucleus disintegrates into a daughter nucleus and an electron (a beta particle). According to the golden rule of energy conservation, the recoiling daughter nucleus and the emitted electron should always share the energy in the same way. As early as 1914, James Chadwick had found something strange about the electrons given off in beta radioactivity – they had a range of energies, rather than a definite value.

Why was there a range of energies when only two particles were produced? Niels Bohr was even prepared to abandon the sacrosanct rule of energy conservation for events at the atomic level.

Then Wolfgang Pauli came to the rescue. Unable to attend a physics meeting in Tübingen, Pauli sent a letter proposing a "desperate remedy" for the radioactivity problem to save energy conservation from the scrapheap. The idea was first heard in public at a meeting of the American Physical Society in Pasadena, California, in June 1931.

The electron given off in beta decay, said Pauli, was accompanied by an invisible particle, which shared the available energy. The particle was almost nothing – just a burst of energy with no electric charge and little or no mass, hardly reacting with matter. Proposing such a radical idea was difficult, even for Pauli, who did not want his lecture printed. However, he was unable to silence a *New York Times* reporter. Several months later, at a major meeting in Rome, Pauli refused again to talk about his new idea in public.

THE LITTLE ONE

Pauli initially called the new particle the "neutron", but when Enrico Fermi was asked in 1932 if this was the same neutron as Chadwick's, he replied "No, Pauli's neutron is very much smaller. It is a neutrino" (meaning "neutral little one"). The joke stuck and became the ghost particle's official name.

Fermi made the neutrino respectable in 1934 when he produced his theory explaining beta-radioactivity. His neutrinos, most likely

WOLFGANG PAULI – "THE SCOURGE OF GOD"

Wolfgang Pauli (1900–1958) is often portrayed as very intolerant. His impatience and sharp tongue helped build up a store of anecdotes. Victor Weisskopf relates how he discovered a mistake in one of his published calculations. Disheartened, he went to Pauli, his teacher, and asked whether he should give up physics.

"Don't," Pauli encouraged. "Everyone makes mistakes. Except me."

Pauli was particularly proud when physicist Paul Ehrenfest called him "the Scourge of God". But those who knew Pauli well say that he simply hated scientific sloppiness and never deliberately set out to hurt anybody. There are also many stories on the so-called "Pauli effect". Every time he walked into a laboratory, an experiment seemed to go wrong.

BROKEN MIRROR

The distinction between left and right is important in everyday life. Many objects are almost left–right symmetric, rotating equally well in either direction, but in practice have a definite "handedness". The Earth's rotation appears right-handed (clockwise) if viewed from the South Pole. Reflected in a giant mirror, the Earth would appear left-handed, spinning in the reverse direction. However, geographical asymmetry immediately distinguishes the real picture from its mirror image.

The basic interactions of particles were initially thought not to distinguish right from left. If a reaction could happen, then so should its mirror image. Most particles spin, twisting around their direction of motion as they fly along. Physicists initially supposed that any particle could equally well spin either clockwise (right-handedly) or anti-clockwise (left-handedly).

Members of the Project Poltergeist team ***Clyde Cowan and Fred Reines (respectively third and fourth from the left) originally planned to catch neutrinos from an atomic bomb explosion. This idea was too far-fetched, and instead they turned their attention to neutrinos from nuclear reactors. In 1953 they saw the first clues that they could be onto a winner.***

having no mass at all, laid the ground for the understanding of a fundamental force of nature – the "weak nuclear force" – which made some nuclei unstable (see page 48). Subsequently the weak force was seen to have an even more fundamental role (see page 66). However, Fermi's revolutionary idea was not accepted for publication in the major publication *Nature* and was printed instead in a relatively obscure Italian journal.

PROJECT POLTERGEIST

Although Fermi's theory established the neutrino, many physicists resigned themselves to it's being an invisible energy-accounting system. Early calculations showed that the ghost particles could travel through many light-years of material without much chance of hitting anything, a daunting assignment for any particle catcher. But in the early 1950s, Clyde Cowan and Fred Reines, two US physicists who had worked at Los Alamos during the war, toyed with the idea of detecting the neutrinos produced during an atomic bomb test.

In trying to dream up a technique to survive the explosion, they realized that an atomic reactor, which also produced neutrinos, would be a more friendly environment. In 1953, they set up their "Project Poltergeist", with tanks containing tons of cadmium solution in front of a nuclear reactor. Counters installed above and below the tanks picked up what looked like the tell-tale signs of neutrinos. But Reines and Cowan could not be sure, and two years later built a larger neutrino catcher at the more powerful reactor at Savannah River, South Carolina.

Neutrinos are reluctant to interact, but the reactor where Reines and Cowan set up their new apparatus produced more than a million million neutrinos per square millimetre (0.01 square inch) per second. With this many neutrinos, their aloofness was finally beaten, and the ten-ton(ne) detector was able to catch about three neutrinos per hour. After a year of careful checks, Reines and Cowan were confident enough to send a telegram to Pauli: "we have definitely detected neutrinos".

With the decays of kaons producing puzzling effects, Tsung-Dao (T.D.) Lee and Chen-Ning (Frank) Yang, two Chinese physicists working at Princeton, had warned in 1956 that mirror symmetry needed to be checked in beta decay. In a simple experiment by Chien-Shiung Wu and her colleagues at Columbia University in New York, nuclear spins were carefully lined up by a magnet. The beta decay electrons, rather than spraying out uniformly, came off on one side.

Many physicists were shocked by the implications. "I do not believe that God is a weak left-hander", said Wolfgang Pauli.

Almost immediately, it was realized that the culprit was the neutrino – involved in beta decay – which can only exist in a left-handed version. The right-handed version of the neutrino is its antimatter counterpart, the antineutrino.

Until the 1930s just two fundamental forces of nature were known: gravity and electromagnetism. Then two new forces entered the arena: a weak nuclear force that could gradually wear down some parts of the nucleus, and a strong force that held the nucleus together.

Nuclear binding glue

THE STRONG AND WEAK NUCLEAR FORCES

HIDEKI YUKAWA – A NATURAL INTROVERT

Hideki Yukawa (1907–81) was born in Tokyo, the son of a geologist. He was timid and quiet as a child, preferring to be alone. In his solitude he read a lot. By the time he entered elementary school he had finished reading a ten-volume edition of *The Chronicles of Taiko,* the tales of an imperial ruler in fifteenth century Japan.

While at school, mathematics was his favourite subject, but he had little interest in physics until Einstein visited Japan in 1922, an event that attracted a lot of popular attention. At that time Yukawa also read a book on recent science which mentioned quantum theory. "I did not understand its meaning at all, but I felt a mystical attraction towards the words", he later recalled.

Just before graduating from Kyoto University in 1929, Yukawa thought of becoming a priest. In his autobiography *Tabibito* (The Traveller), he wrote: "A dislike for society resides within me even today, although it is more a desire to avoid contact than an active dislike. I want my interactions with other people cut down to about one-tenth..."

With the discovery of the neutron in 1932, the composition of the atomic nucleus seemed to be understood in terms of protons and neutrons. But, as often happens in science, no sooner is one problem solved than another appears. Protons, all having the same electric charge, repel each other violently, while neutrons, since they have no charge at all, cannot counteract this repulsion. With only electromagnetism at work, all nuclei ought to explode!

But physicists knew that nuclei are difficult to break apart. What held protons and neutrons together? Whatever it was, it had to be very powerful. Electromagnetic repulsion between protons increases as they get closer together. In the tiny nucleus, the protons and neutrons had to "feel" a force stronger than anything previously encountered in physics. Gravity, the only other known force, held planets in their orbits, but was much too feeble to play any role on the atomic scale.

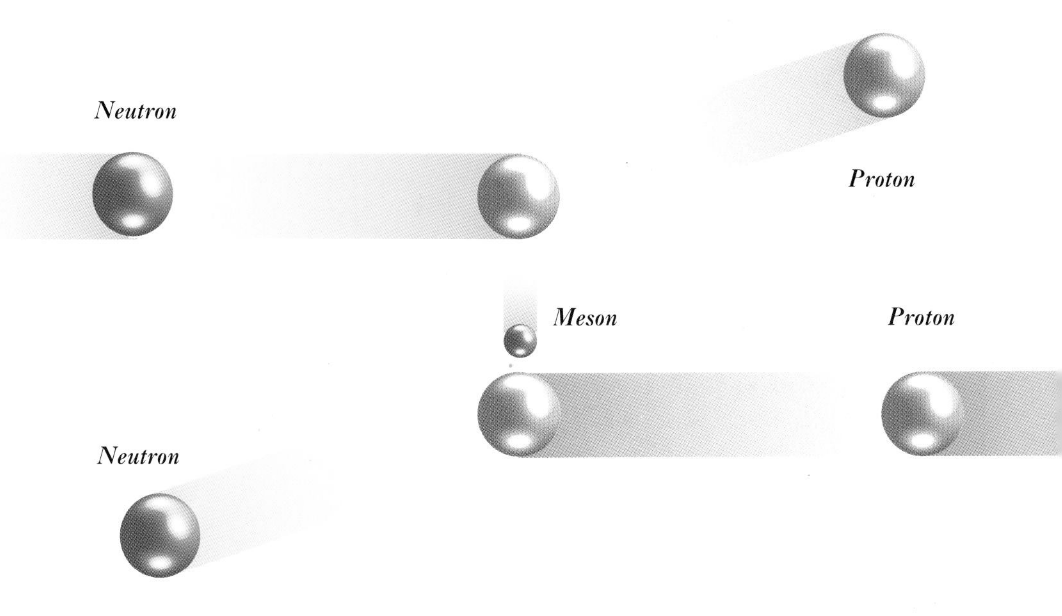

Messengers of force *In the quantum world, energy can be borrowed free of charge as long as it is paid back sufficiently quickly – before nature has time to "notice". This borrowed energy can appear in the form of a particle, which briefly flickers into existence. These transient particles are the messengers of physics, transmitting an effect from one particle to another. This is how a force works. A neutron encountering a proton will interact with it by the exchange of a smaller electrically charged meson, causing the proton and neutron to swap charges and identities. Nuclei are held together by the constant swapping of mesons between protons and neutrons.*

ERNEST LAWRENCE AND HIS MACHINES

Ernest Orlando Lawrence (1901–58) was born in Canton, South Dakota. His major interest was electricity. By the age of nine he had a cellarful of electric motors and spark coils. His constant ambition worried his mother: "You don't have to go so fast. There is plenty of time." This frantic energy and enthusiasm became hallmarks of Lawrence's work. The photograph shows him (white shirt, kneeling) and members of his team with their 27-inch (69-centimetre) cyclotron in 1932. The third of Lawrence's cyclotrons, it accelerated protons to 4.8 MeV. During the depression of the 1930s, Lawrence had to fight hard for finance. One windfall, an old electromagnet from the Federal Telegraph Company, was used in several cyclotrons. Shortage of funds may be why Lawrence's laboratory missed the big discoveries of the 1930s.

NUCLEAR BENEFITS

During the 1930s, bigger and more powerful cyclotrons were built at Berkeley and became important tools in many branches of science. One area of particular interest to Lawrence was the production of radioactive isotopes for medicine and biology, where he collaborated with his brother John, director of Berkeley's medical physics laboratory. In 1938 they also used an accelerator for cancer therapy. High-energy beams destroy matter and are dangerous, but their idea was to use a narrow beam and irradiate only the cancerous tissue. This technique has become a major weapon against cancer.

By 1939 the biggest cyclotron was 1.5 metres (5 feet) in diameter and reached19 MeV. To produce Yukawa's particle, Lawrence planned to build a 4.6-metre (15-foot) machine, which would have been capable of 340 MeV. This project was interrupted by World War II, but in any case the cyclotron technique had run up against a technical problem that limited its maximum energy to about 20 MeV. As the particles approached the speed of light, Einsteinian relativity made them fall out of step with the accelerating electric field. Higher energies had to wait for the development of ingenious new techniques.

Enter the muon

A SURFEIT OF PARTICLES

In 1936, four years after he discovered the positron, Carl Anderson found a new particle in cosmic rays. Today called the muon, it was completely unexpected and did not fit in with the atomic theories of the time. "Who ordered that?" demanded US physicist Isidor Rabi on hearing the news.

After his discovery of the positron in 1932, Carl Anderson and his assistant Seth Neddermeyer continued with cosmic ray experiments. Most of the cosmic particles at sea-level were highly penetrating, able to travel through thick layers of lead. What were they? Anderson for a time thought they could be electrons. He explained certain peculiarities in their behaviour by speaking of "green" and "red" electrons.

But it was becoming increasingly clear that the findings did not fit in with the known subatomic picture. In experiments at Pikes Peak in Colorado in 1935, Anderson found the first evidence that he was dealing with an entirely new kind of particle. A year later, on 12 November 1936, Anderson and Neddermeyer were confident enough to go public. That year Anderson received the Nobel Prize for his positron discovery.

Cosmic particles ***Carl Anderson inspects his magnetic cloud chamber. Muons are produced when particles from space, mostly protons, hit atoms and nuclei in the upper atmosphere. These collisions create unstable pions, which within much less than a millionth of a second turn into muons and neutrinos that shower down to the Earth's surface. Muons leave characteristic tracks in a detector such as a cloud chamber.***

ANDERSON'S BOMBER MISSION

Carl Anderson, born in New York City in 1905 and an accomplished high jumper in his youth, earned a physics Ph.D. at Caltech (California Institute of Technology) in Pasadena in 1930. For his research thesis he studied the photoelectrons that are produced by X-rays, but later wanted to learn more about quantum mechanics. Robert Millikan, his professor, asked him instead to build an instrument that could be used to measure the energies of electrons in cosmic radiation. It was good advice. With this apparatus, a cloud chamber with a magnet, Anderson discovered the positron in 1932 (see page 44) and later the muon.

During World War II, Anderson worked on artillery rocket research and in 1944 spent a month on the Normandy beaches with the Allied troops. This brought him good military contacts, and after the war he was able to obtain a B-29 "Superfortress" bomber for his cosmic ray research. The plane was modified for extreme altitudes and made in all 35 flights, each of about 5 hours, carrying a cloud chamber. The highest altitude was 12,000 metres (41,000 feet). In the thin atmosphere at these altitudes, cosmic rays are several hundred times more abundant than they are at ground level.

NAME GAME

Anderson first called the new particle "mesoton" – from the Greek "meso" meaning "in between". But Robert Millikan, the leading US cosmic ray expert, who had been away when the announcement was first made, insisted the spelling should be changed to "mesotron", with an "r", as in electron. Subsequently the name was contracted to "meson", the name of the nuclear force carrier that Yukawa had predicted in 1935. Many physicists thought that Anderson's meson might be Yukawa's particle because it had the correct mass.

Anderson, however, was convinced from the outset that his particle and Yukawa's meson were not the same. His particle seemed to be immune to nuclear forces, quite unlike Yukawa's particle, which had to interact strongly.

World War II interrupted the search, but Yukawa's particle was finally found in 1947, and the question resolved. Yukawa's particle was the pi-meson, or pion (see page 44), while Anderson's particle was called the mu-meson, or muon. The muon was not properly identified until the 1950s. It is a heavy brother of the electron, 205 times heavier in fact.

In 1943, three Italian physicists, Marcello Conversi, Oreste Piccioni and Ettore Pancini were trying to study the new cosmic-ray meson in their Rome laboratory. When their part of the city was bombed by the Americans, they had to move to a safer place. Soon afterwards Rome was occupied by German troops. Nevertheless they were able to complete their experiment, showing that the particle passed relatively easily through all kinds of nuclear matter. The cosmic-ray meson could not be the same as Yukawa's meson, which would have to interact readily with nuclei.

Move fast, live long *Muons are unstable and at rest decay after about 2-millionths of a second. At first glance, therefore, muons in cosmic rays should penetrate only about 600 m (2,000 ft) of Earth's atmosphere before breaking down. But some penetrate the 10 km (6 miles) to sea-level. This is due to an effect called "time-dilation", which follows from Einstein's special theory of relativity. Time runs slower for objects with a speed close to that of light. From the muons' point of view nothing strange happens, but when seen from the Earth it appears that a fast-moving muon lives many times longer than one at rest.*

A strange family of geriatrics

PARTICLES WITH STRANGENESS

In the 1950s a new family of peculiar, unstable particles was found. Having just sorted out the muon and the pion, physicists now had to cope with kaons, lambda, sigma and ksi particles. The new species was dubbed "strange particles" because they lived much longer than anybody expected.

In 1947 George Rochester and Clifford Butler of Manchester University spotted an unusual effect in cosmic rays. In their cloud chamber, two tracks emerged from a point, giving an inverted V shape. The researchers concluded they were seeing an unknown particle decaying into two secondary particles.

In 1950 Carl Anderson confirmed the discovery from 11,000 cloud chamber photographs taken on the top of White Mountain, California, where at 3,000 metres (10,000 feet) there were about 40 times more cosmic rays than at sea-level. He found 34 examples of the new particle, now called the "K-meson" or "kaon".

Eager to find out more about the new particle, physicists soon found that kaons decayed in a peculiar way. A useful time interval for nuclear physics is 10^{-23} second, roughly the time it takes for a light ray to cross a nucleus. From the way kaons were produced, at first sight they should only live for about one such "nuclear year". In fact they lived for about 10^{-8} second, an astonishing 1,000 million million times longer. Because of this slow death, kaons were dubbed "strange particles".

In the early 1950s, the pioneer cosmic-ray discoveries of strange particles were confirmed and extended by a new generation of experiments using high-energy machines. Soon three more particles were discovered, called "lambda", "sigma" and "ksi". While kaons were several times heavier than pions but still lighter than the proton, these additional strange particles were the first examples of a new breed of particle even heavier than the proton but too unstable to live in natural nuclei.

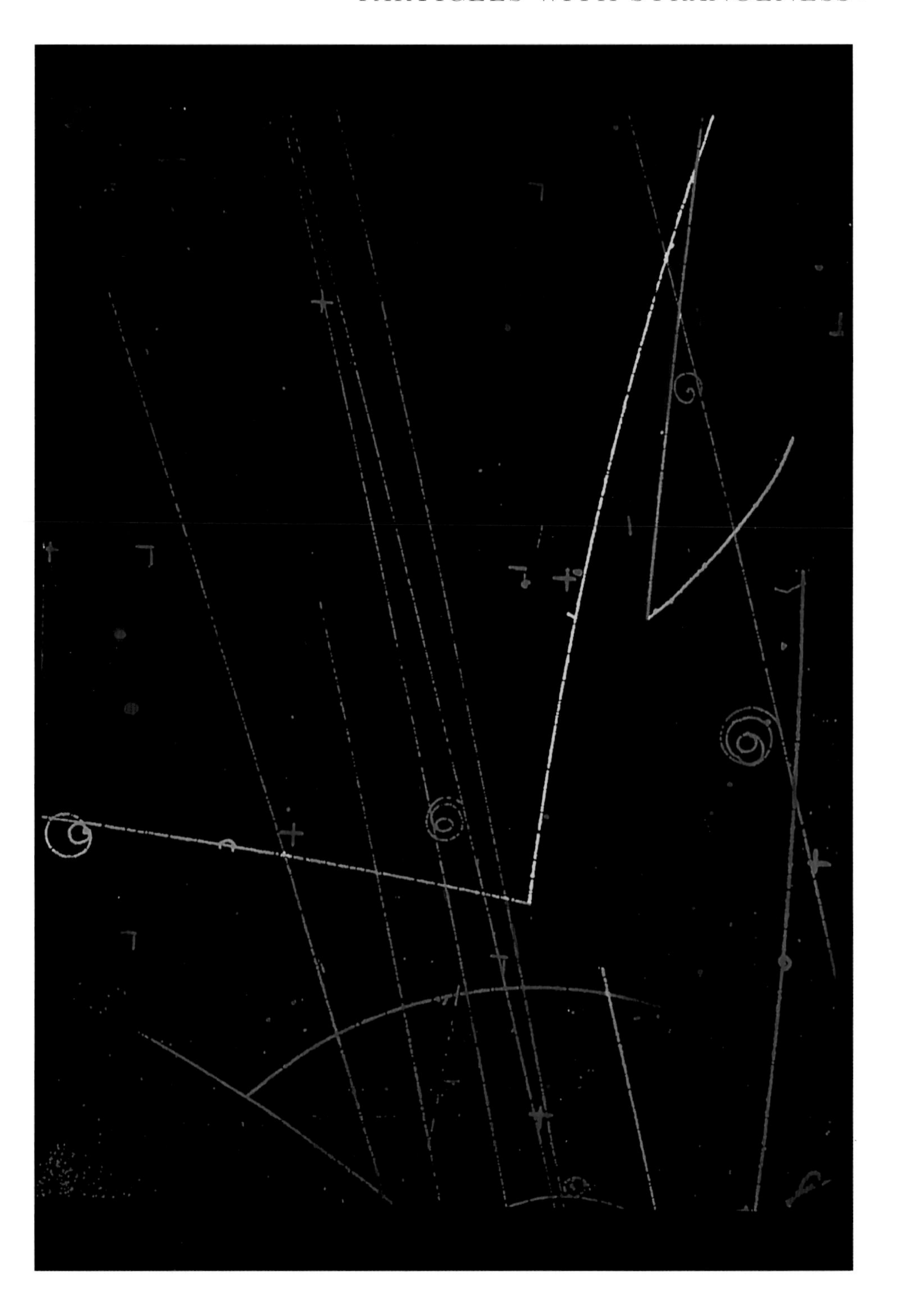

BUBBLE AWAY

Progress in physics goes hand in hand with new methods of detection, the classic example being the invention of the telescope and microscope in the sixteenth and seventeenth centuries, which for the first time extended the power of the human eye.

In 1952 a breakthrough came when the American physicist Donald Glaser invented a new instrument for detecting particles, the bubble chamber. This device replaced the cloud chamber, which was too inefficient for detecting higher-energy particles – they just whizzed through the water vapour in the chamber without leaving any tracks.

Instead of water vapour, the new device was filled with a superheated liquid – one slightly above its normal boiling point but kept under pressure to stop it from boiling. This pressure was suddenly reduced as particles entered the chamber, and bubbles immediately formed in the particles' wake.

Glaser had got the idea from watching bubbles forming in a bottle of beer when the pressure suddenly dropped as the cap was taken off. His original bubble chamber, a glass bulb 2 centimetres (about 1 inch) in diameter filled with diethyl ether (illustrated here), could record the tracks of cosmic rays. Jack Steinberger was the first physicist to make use of a bubble chamber. His experiments at Brookhaven in 1954 discovered the neutral sigma particle and showed how useful the new technique was for seeing the short tracks of particles that decayed within a fraction of a second.

Meanwhile Luis Alvarez at Berkeley had also heard about Glaser's invention and immediately set to work to build a large bubble chamber filled with liquid hydrogen. He also had the idea of synchronizing the chamber with the pulses of particles fed by the accelerator.

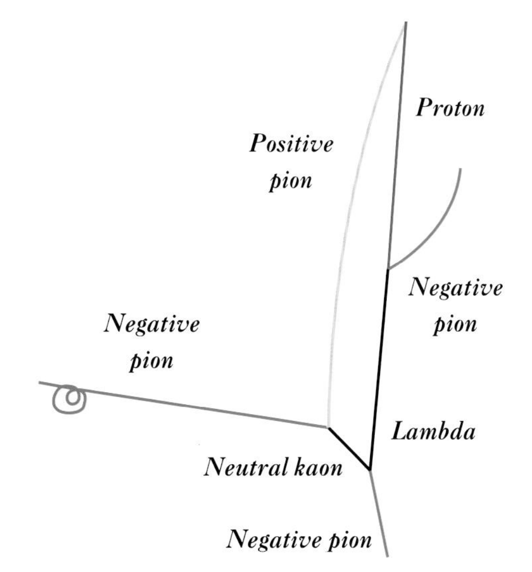

Strange particles (left and right) *A false-colour bubble-chamber photograph showing the presence of particles containing strange quarks, and a diagram of the particle events. A negative pion enters from below and interacts with a proton in the liquid filling the bubble chamber, producing two strange particles: a lambda and a neutral kaon. Because they are both neutral, these particles leave no tracks, but they show their presence when they decay. The lambda turns into a proton and a negative pion; the kaon into a positive pion and a negative pion. The blue tracks in the photograph are particles not involved with the interaction.*

STRANGENESS

In 1954 Murray Gell-Mann in the United States and Kazuhito Nishijima in Japan explained the relative longevity of the new particles. As well as electric charge, the new strange particles carry another fundamental property, called "strangeness".

Like electric charge, strangeness has to be conserved. In the same way that electric charge is shuffled around but never lost, so the total amount of strangeness must be the same before and after a reaction under the strong nuclear force takes place. Under this precise strangeness accounting, the lightest strange particles cannot decay under the strong nuclear force. Decaying instead through the far more leisurely weak nuclear force, they manage to survive for a relatively long time.

A PARTICLE ZOO

By the end of the 1950s, the simplicity of a particle world once confined to protons, neutrons and electrons had been replaced by a bewildering profusion of new unstable particles. Including positive, negative and neutral electric charge versions of particles, there were three pions, three kaons, three sigmas, two ksis, two muons and a lambda.

This particle "zoo" resembled the confusion in chemistry a hundred years earlier, when only a totally new classification scheme could bring order into the disarray of an ever-growing list of elements. Why were there so many unstable particles? They played no role in ordinary matter, and disintegrated into stable particles. Did this bewildering particle profusion hint at a simpler substructure deep inside?

All the new post-World War II particles felt the strong force and could be grouped with the proton and neutron as hadrons. These could be subdivided into mesons (particles such as pions and kaons, heavier than electrons but lighter than protons) and the heavy baryons (the proton, the neutron and their still heavier relatives such as the lambda and the two ksis).

The first big machines

SYNCHROTRONS AND THE COLD WAR

In the search for more new particles in the 1950s, bigger and more powerful particle accelerators were built. The energies found in cosmic rays now could be reproduced for the first time, and under carefully controlled conditions. Physicists could chose from a "menu" of beams.

After World War II a new particle accelerator – the synchrotron – was invented by Edwin M. McMillan, a colleague of Ernest Lawrence. In this machine the particles were held in a circular orbit inside an evacuated tube instead of spiralling out as they did in the cyclotron. The huge enveloping circular magnet of the cyclotron was replaced by a ring of smaller C-shaped magnets.

Particles could now be accelerated to velocities close to the speed of light. Both the electric field accelerating the particles and the magnetic field bending their path were synchronized with the steadily increasing energies to take into account the subtle effects of Einstein's special theory of relativity.

In 1952 the first proton synchrotron, at Brookhaven National Laboratory near New York, reached 3,000 million electronvolts (3 gigaelectronvolts, or GeV). The first accelerator to produce energies within the range of cosmic rays, it became known as the Cosmotron. This 2,000-ton(ne) machine led the field until a 6 GeV synchrotron, the Bevatron, went into operation at Berkeley in 1954. In 1955 this machine attained its goal of producing the first anti-protons, the antimatter counterparts of protons.

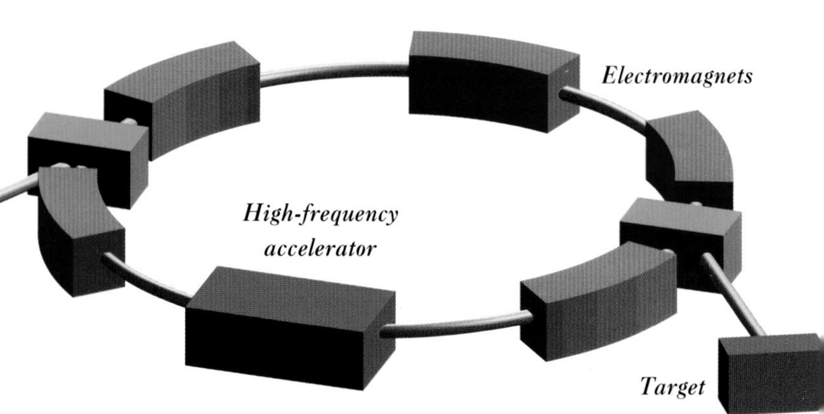

The synchrotron (above) *Particles are injected into a ring of electromagnets, which hold them in a circular orbit. High-frequency electric power accelerates the particles. Both the accelerating power and the magnetic field are synchronized to maintain the particles in orbit, before they are finally ejected towards their target.*

FROM THE RUINS OF WAR

In December 1949 a United Nations cultural conference in Lausanne, Switzerland, looked at the sorry state of post-war European science. The Old Continent lay in ruins, scientists had been dispersed and the physics pendulum had swung towards to the United States. The continent that had nurtured basic physics for two thousand years, from the ancient Greeks to Rutherford, Bohr, Dirac, Heisenberg and Pauli in the twentieth century looked like losing its cultural heritage to the New World.

At the Lausanne meeting, a message from the eminent French physicist Louis de Broglie recommended setting up an international research laboratory. Six months later a UNESCO (United Nations Educational, Scientific and Cultural Organization) resolution was passed, and what came to be known as CERN (Conseil Européen pour la Recherche Nucléaire, later the European Laboratory for Nuclear Research), arose from the ashes of the war. European cities vied to be home to the new laboratory, but the village of Meyrin, just outside the Swiss city of Geneva, was finally chosen. Building work started in the summer of 1954.

COLD WAR, WARM COMPETITION

CERN's main objective was to build a 28 GeV proton synchrotron, the world's biggest particle accelerator, with a circumference of 600 metres (2,000 feet). After several fanciful suggestions, it was modestly christened "PS", for proton synchrotron.

In 1957 the Soviet Union unveiled a "synchrophasotron" at Dubna, just outside Moscow, which reached a new world record of 10 GeV. Meanwhile Brookhaven National Laboratory in the United States was planning its new AGS (Alternating Gradient Synchrotron), aiming for 30 GeV. CERN and Brookhaven were out to break the Russian record and on 24 November 1959 the Europeans won the race when the PS came on line.

The new Brookhaven machine was initially proposed as a logical scale-up of existing techniques. However, knowing that the European team were breathing down their necks, the Brookhaven team came up with the new "alternating gradient" idea, which made for more economical magnets. The CERN team also adopted the new idea and were the first to get it to work with protons. However, the Brookhaven machine was quicker into action for physics research and was soon the scene of major discoveries. The Europeans had to wait ten years before reaping a comparable research harvest.

Soviet synchrotron (below) *Called the* synchrophasotron *by the Soviets, the 10 GeV synchrotron at the Dubna Laboratory, Moscow started operation in 1957 and was for two years the world's highest-energy accelerator, until 1959 when CERN's Proton Synchrotron wrested away this honour.*

"NOT TO BE OPENED..."

On November 24, 1959 the Europeans took the world record for particle-beam energy from the Soviet Union, and a bottle of vodka, especially sent by the Soviets for the occasion, was passed around the CERN control room. The team leader of the CERN proton synchrotron, John Adams, put documentary proof of the achievement in the empty bottle before sending it back to Moscow.

John Adams, a tall, athletic-looking Briton personified the creation of CERN, where hundreds of people with different backgrounds and from countries, which only a few years earlier had been at war, worked together towards a common goal.

Born in 1920, Adams had no standard academic training. With no money for further education, he went to work as an apprentice in an electrical engineering firm and was about to be promoted when the factory was bombed during World War II. Too young to be called up, Adams went into the Government's telecommunications research laboratory to work on radar. After the war, he turned to a series of new research challenges at the Harwell Laboratory.

He had an impressive ability to inspire and motivate people, and scientists and engineers respected him also for his intelligence and alertness to problems, while mechanics appreciated his ability to talk to them in their own language. The headquarters of the CERN Proton Synchrotron group became known as "Adams Hall". After three separate spells as Director General of CERN, John Adams died in 1984.

The second ghost

DISCOVERY OF THE MUON-NEUTRINO

The discovery of the elusive neutrino in 1953 was a major breakthrough. But the break-up of cosmic ray muons suggested that this wraith-like particle might also come in different kinds. In the early 1960s another classic discovery found a second neutrino to partner the first.

Jack Steinberger, a Jewish refugee from Nazi Germany, began his pre-World War II studies in the United States in chemical engineering. After the war, he switched to physics at the University of Chicago. One of his teachers was Enrico Fermi, who had shown how the weak force could be understood. Fermi's courses were gems of simplicity and clarity, Steinberger recalls.

After being asked by Fermi to look into a problem with cosmic-ray muons, in 1948 Steinberger discovered that these muons break up into three particles, probably one electron and two neutrinos. To get the

Spark chamber *Melvin Schwartz with the Brookhaven spark chamber, used to detect the first muon-neutrinos in 1961–2. The use of "synthetic" neutrinos from the decay of particles in a beam became a major research tool. The muon-neutrino discovery was the first of many major breakthroughs.*

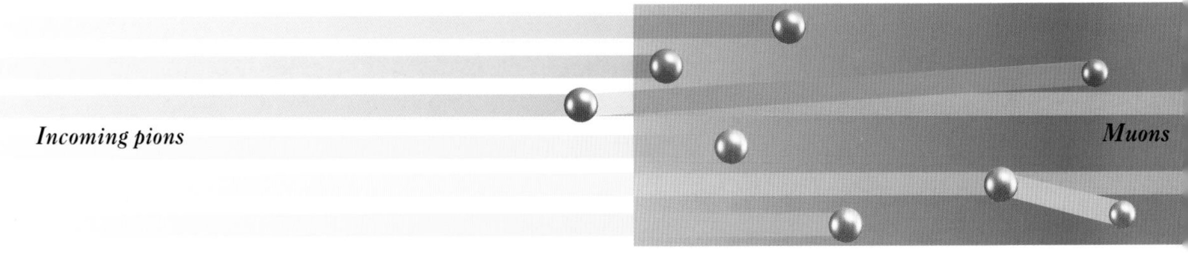

accounting straight, the two "neutrinos" had to be a neutrino and an antineutrino. Normally a particle and its antiparticle annihilate, disappearing in a burst of energy. As nothing of the kind was seen, some physicists suspected the two neutrinos were somehow different.

Steinberger, Leon Lederman and Melvin Schwartz hoped that they could learn more about the weak force by making the new proton synchrotron at Brookhaven National Laboratory produce a beam of high-energy neutrinos.

Nobody had produced a neutrino beam before, and to achieve it needed some hefty engineering. The Brookhaven accelerator's protons, smashing into a beryllium target, produced a beam containing a variety of particles, including a lot of very fast pions. Left alone, these disintegrated after travelling 20 metres (65 feet) into muons and neutrinos. Then came the crunch: 13.5 metres (45 feet) of steel armour plate from a naval scrapyard. Built to withstand heavy bombardment of another kind, the plates stopped all particle projectiles in the beam except neutrinos, most of which passed clean through.

Behind the steel wall was a new kind of detector with nearly a hundred 2.5-centimetre (1-inch) thick aluminium plates, together weighing 10 ton(ne)s. The physicists hoped that from time to time a neutrino would bump into an aluminium nucleus, producing either muons or electrons. With a high voltage across the gaps between the plates, these particles would produce sparks.

Over eight months in 1961–2 there were just 25 days when everything worked perfectly. Out of 100 million million neutrinos passing through the detector, 51 reactions were recorded. But every time it was a muon, rather than an electron, that was created. The neutrinos from the pion beam clearly remembered their parentage – born in decays producing muons, they went on to interact and produce muons. So they were a different kind of neutrino: a muon-neutrino.

JACK STEINBERGER – AN EYE FOR NEW TECHNIQUES

Jack Steinberger was born in Germany in 1921, the son of a religious teacher in the local Jewish community. To safeguard their future, in 1933 Jack's father sent his sons to the United States. The rest of the family followed in 1938. During World War II, Jack joined the army and was sent to a laboratory developing radar bomb sights, where he was introduced to physics.

After his initial career in the United States, Jack Steinberger returned to Europe, and has worked at CERN since 1968. There he was one of the first to use a new technique invented by Georges Charpak that enabled detectors to be connected directly to electronics and computers.

THE LIGHT ONES

With the discovery of the muon-neutrino, four fundamental particles were known that did not feel the strong nuclear force. The electron and the muon, each paired with its respective neutrino, were grouped together as leptons, from the Greek word for "small" or "delicate". Leptons have the weak nuclear force in common. Today the family has a third pair, the tau and its neutrino. The tau particle, found in 1975, is 3,500 times heavier than the electron. Despite fears of a ladder of successively heavier leptons, physicists are now confident that there are no more to discover (see page 78).

Steinberger, Lederman and Schwartz received the Nobel Prize in physics in 1988. Having "waited" 27 years for the reward, Steinberger remarked, "To get the Nobel Prize in physics you have to do two things. Come up with an interesting experiment while you are young and then stay alive long enough."

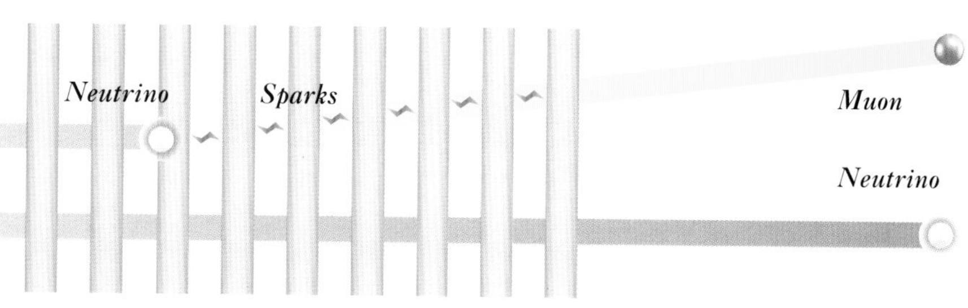

A trap for neutrinos *In the neutrino-beam experiment, 13.5 m (45 ft) of steel shielding filtered off everything produced by the incoming pions except the neutrinos. It did not matter if a few neutrinos failed to make it. But the detector behind the steel shielding had to be big enough to catch some of them. Reminiscing on the experiment, Mel Schwartz relates "in planning for experiments, I tend to be an optimist. When we first sat down to do the figures, we said 'well, we ought to get one event per ton per day'. That was the number we worked with. It turned out to be smaller than that, but fortunately we built a ten-ton detector".*

Adventures in quarkland

PROPOSALS FOR A NEW LAYER OF MATTER

By the early 1960s, physicists were on the verge of despair. Instead of having just a few basic particles to contend with, there was a subatomic glut. The theorists came to the rescue. All of the new particles could be explained by just three hypothetical units, whimsically named "quarks".

Worried about having so many particles to contend with, Murray Gell-Mann of Caltech and the Israeli physicist Yuval Ne'eman in 1961 applied ideas of mathematical symmetry to the thirty-or-so strongly interacting particles then known, many carrying the new "strangeness" label. Independently they discovered how the different particles, grouped into families containing eight, or sometimes ten, members, fitted into special geometrical patterns. Gell-Mann called the scheme the "Eightfold Way", an expression borrowed from Buddhism.

It was a sort of subatomic periodic system, parallelling Mendeleyev's famous table of chemical elements of a century earlier. Gell-Mann also echoed Mendeleyev's successful prediction of new elements. At a physics conference at CERN in 1962 he dramatically predicted a new particle – the omega minus – to fill a gap in a family of ten. He predicted its exact properties and in December 1963 a team at Brookhaven managed to detect one in their bubble chamber.

At first the pattern makers had not been taken seriously. But the discovery of the omega minus made everybody sit up and take notice. What made the symmetry idea work? What lay behind the neat eight- and ten-fold patterns? Working independently of each other, Gell-Mann and his Russian-born Caltech colleague George Zweig showed that the families followed naturally from the different ways of mathematically arranging three basic constituents.

Gell-Mann first adopted the word "quork". Some people thought that he was saying "quirk", but anyway later Gell-Mann came across "quark" – "Three quarks for Muster Mark!" – in James Joyce's book *Finnegans Wake*. Published in 1939, this difficult novel is a laboratory for experiments in language. Reluctant to risk having his revolutionary quark idea refused by a staid journal, Gell-Mann instead submitted his paper to a relatively new European publication early in 1964. The references include Joyce's book. Working at CERN on a year's leave from Caltech, Zweig chose the name "aces", but his ideas never got into print at the time.

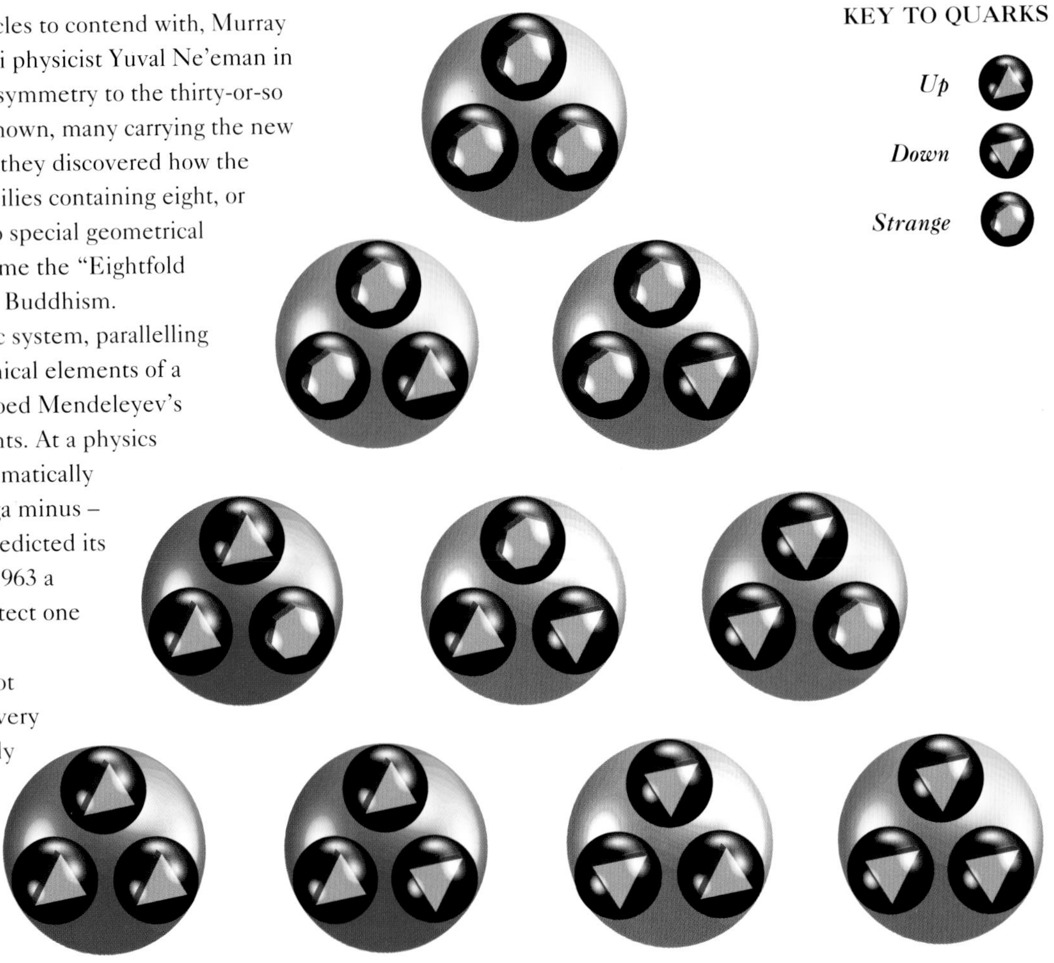

KEY TO QUARKS

Up

Down

Strange

Buddhism predicts the omega minus *One of the most attractive features of quark theory was how subatomic particles could be arranged into neat geometric patterns. Inspired by the Buddhist Eightfold Way, Gell-Mann constructed a pyramid pattern containing heavier relatives of the proton and neutron and their strangeness-carrying counterparts. At the base of the pyramid came four particles, heavier versions of the proton-neutron family and carrying no strangeness. After two rows of particles, sucessively containing one and two strange quarks, the top of the pyramid needs to be filled by a single particle carrying a full house of three strange quarks. This is the omega-minus, whose discovery in 1964 dramatically confirmed Gell-Mann's idea.*

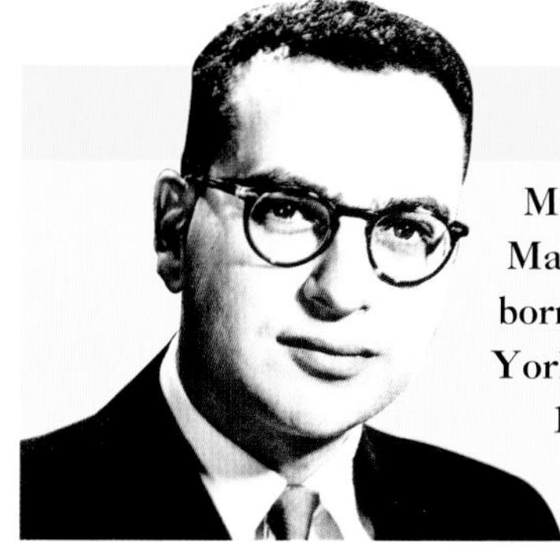

GELL-MANN – ATTRACTED BY COMPLEXITY

Murray Gell-Mann was born in New York City in 1929, the son of Austrian immigrants. As a teenager his early interests were in archaeology. As well as the strangeness and quark ideas, he has made several other major contributions to physics. A lifelong hobby is also linguistics, the origin and evolution of words in different languages. Always attracted by intellectual challenges, in 1984 he helped set up a research centre in Santa Fe, New Mexico, devoted to "complexity" – systems that are inherently difficult to describe.

FRACTIONS AND COLOUR

Because of the way quarks fitted together, it was natural for Gell-Mann to label them by directions. Two were called "up" (or "u") and "down" (d), while the third quark was named "strange" (s) because it was a vital component of strange particles.

The quarks had to carry electric charge, but unlike other particles quark charges appeared to come in fractions of the charge on the proton, the up quark with $+\frac{2}{3}$ and the down quark $-\frac{1}{3}$. Physicists had never seen fractional charge and many refused to believe in it. But nevertheless, simple quark rules accounted for all known hadrons. The baryons are quark triplets, a proton having two up quarks and one down quark, and a neutron having two down quarks and one up. Mesons are quark–antiquark pairs.

The idea of a new layer of matter carrying split electric charges was difficult to accept, so the quark idea was first sold as a purely mathematical scheme which somehow made everything come out right. Quarks have no independent existence and are not real constituents, many physicists said.

Besides split electric charges, something else discouraged the idea that quarks were real. The newly discovered omega minus, although a triumph for the quark theory, brought with it a dilemma. According to a basic quantum rule – the Pauli exclusion principle – no two quark-like particles can be in the same state inside some larger particle. Yet the omega minus consisted of three apparently identical strange quarks.

To sidestep the problem, O.W. ("Wally") Greenberg suggested in 1964 that quarks carried an additional charge, known as "colour". Colour distinguished the otherwise identical quarks from each other and saved the exclusion principle. The three types of quarks each come in three different colours – usually referred to as red, green and blue – and a quark triplet always must have one quark of each colour, giving a colour-neutral or "white" result, analogous to the mixing of coloured light. (Apart from this neat analogy, "colour" in the quark world has no relation to the everyday use of the word.)

With the quark idea on firmer ground, physicists found that this astonishingly simple picture could explain many aspects of particles. At first restricted to static properties – particle "labels" such as electric charge and strangeness – it was gradually extended to dynamics, comparing the rates of different particle reactions. Baryons, with three quarks, offer more reaction possibilities than mesons, with just two.

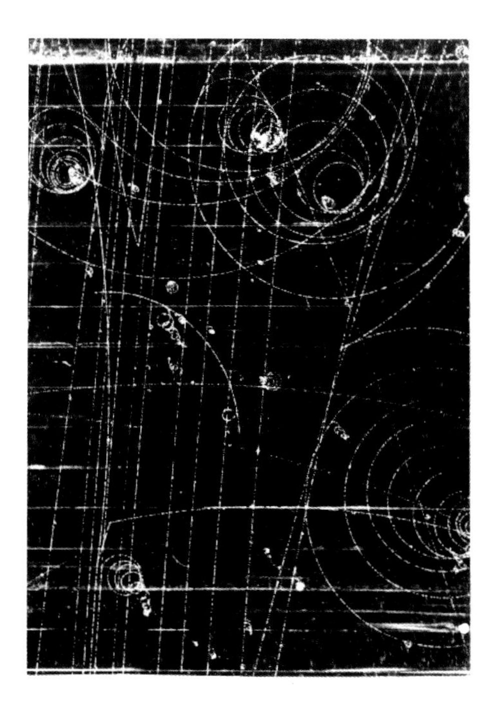

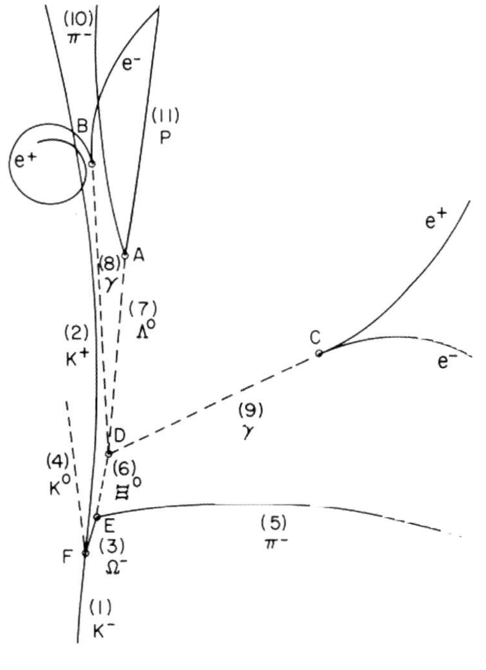

Omega minus *In the audience when Gell-Mann made his 1962 prediction of the omega-minus was Nick Samios from Brookhaven, where a big new bubble chamber was being completed. Gell-Mann's particle carried more strangeness than any other particle, and Samios realized this strangeness would have to be supplied by a beam of particles with strangeness: kaons. The experiment had only just begun when some reflectors (nicknamed "coat hangers") installed to improve the view through the bubble chamber's window broke loose and threatened to wreck the chamber. However, the order was given to carry on. Four weeks and 50,000 photographs later Samios' team spotted a picture-postcard example of an omega-minus decay, fanning out from the bottom left of the photograph.*

The hunting of the quark

PROBING INSIDE THE PROTON

Near Stanford University in California, Route 280 from San Jose to San Francisco crosses over a 2 kilometre (2-mile) concrete gallery, the Stanford Linear Accelerator Center. In the late 1960s, this powerful machine was the scene of an experiment that showed that quarks exist.

Despite its attractive simplicity, the quark picture was not ready for the textbooks. As well as their awkward fractional charge, quarks stubbornly refused to come out in the open. Experimenters searched for them in cosmic rays, meteorites and sludge from the seabed, but in vain. When high-energy protons were smashed into a target, no quarks came flying out. The official explanation was that quarks were not real objects, but mathematical ideas that fitted the pattern of the Eightfold Way. Nevertheless back-of-the-envelope calculations using simple quark counting continued to give compelling results.

Under the influence of Ernest Lawrence, physicists at the Berkeley Laboratory north-east of San Francisco had invested heavily in large circular machines, and most of the rest of the world followed suit. However, at Stanford University, to the south of San Francisco, electron specialists had other ideas. They concentrated instead on using high-frequency oscillating electric fields to accelerate particles. In 1937, a major breakthrough was the invention of the klystron, a powerful new kind of radio-frequency amplifier.

When research resumed after World War II, the new techniques were soon put to work at Stanford. Robert Hofstadter used powerful new electron beams to

Physics landmark ***Alongside the Stanford University campus at Palo Alto, California, the 3-km (2-mile) gallery of the Stanford Linear Accelerator Center (SLAC) is an unmistakeable feature of the landscape. It is now officially classified as a US National Historic Engineering Landmark.***

Quarks in particles *Quarks bind together in threes to form baryons – the proton and neutron and their heavier nuclear cousins. For example, the proton (positive charge) is made up of two up quarks (each of charge $+\frac{2}{3}$) and a down quark (charge $-\frac{1}{3}$). Quarks can also bind with their antimatter counterparts, antiquarks, to form mesons, such as the pion.*

BARYONS

Proton

Neutron

MESONS

Positive pion

Negative pion

RICHARD FEYNMAN – IRREVERENT GENIUS

Like Einstein, Richard Feynman was a legend in his own time. Seldom able to take anybody's word for anything, he preferred to think each new problem out for himself. In doing so he would often discover new ways of looking at physics. For instance, his solution to quantum electrodynamics, based on simple diagrams where one dimension represented space and the other time, provided a new method of depicting nature's forces. Julian Schwinger, another architect of quantum electrodynamics, said that "just like today's silicon chip, Feynman diagrams brought calculations to the masses." Few could match Feynman's flamboyance and outspokeness. His anecdotal auto-biography, *Surely You're Joking Mr. Feynman,* published in 1985, was an immediate best-seller.

His final major contribution came just before his death in 1988, when he was asked to serve on the official commission that investigated the 1986 Cape Kennedy disaster, in which the *Challenger* Space Shuttle exploded shortly after it had taken off. His demonstration of the effect of low temperatures on the *Challenger*'s sealing gaskets using a cup of iced drinking water was typical of his remarkable ability to explain complex problems in the simplest terms.

show that the proton was not just a fuzzy round ball. Inside the ball, some areas were less fuzzy than others: the proton had a definite shape. Meanwhile a gifted band of physicist-engineers, including Wolfgang ("Pief") Panofsky, planned a "monster" electron machine 3 kilometres (2 miles) long. With cash easy to come by in those days, it was soon built, and the Stanford Linear Accelerator Center (SLAC) delivered its first beams in 1967.

One of the first experiments to use SLAC was led by Jerome Friedman, Henry Kendall and Richard Taylor. Their initial idea was to check whether the new particles then being seen at other machines could also be produced by electrons. However, a theorist, James Bjorken, had been toying with some new ideas and suggested they might also look for something else.

Following Bjorken's advice, they swung their huge detectors away from the line of the electron beam to look for electrons which had swerved violently. They were surprised to see ten times more wide-emerging electrons than expected.

RUTHERFORD REVISITED

It was a rerun of Rutherford's 1911 classic experiment (see page 32). He had seen his alpha particles bouncing back from heavy, dense nuclei hidden deep at the centre of atoms. Richard Feynman, passing through SLAC, put his penetrating insight to work, simplifying Bjorken's abstract ideas and showing how the results could be explained by assuming that protons contained a few tiny hard grains, which he called "partons".

The experimental results were first announced at a major scientific meeting in Vienna in August 1968, where the conclusion was that they "might give evidence on the behaviour of point-like charged structures within the nucleon". At first the message fell on infertile ground. It took a year before the parton idea began to sell. But few physicists dared yet to speak of quarks.

Now that physicists knew there was something deeper to the proton, the next task was to look hard at the new partons. By 1974, neutrino beam experiments at CERN had shown that Feynman's partons had fractional charge, and there were three of them inside each proton. Gell-Mann's mathematical quirks were for real.

Henry Kendall, Jerome Friedman, both now at Massachusetts Institute of Technology, and Canadian Richard Taylor, still working at SLAC, received the Nobel Prize in physics in 1990, over 20 years after the experiment that discovered the quark.

Einstein's last dream

THE GRAND UNIFIED THEORY

For most of his life, Albert Einstein strived to build an ultimate theory of the Universe, unifying electromagnetism and gravity. He worked on this project relentlessly from the 1920s until he died in 1955. But his efforts were doomed to fail because he ignored nuclear forces.

In 1915 Albert Einstein's general theory of relativity gave a new description of gravity, replacing Newton's old theory. According to Einstein, gravity is not a force in the conventional sense, but a geometrical property of space caused by the matter in it: space becomes curved near heavy objects. He predicted that light rays from a star that passed behind the Sun would be bent by the Sun's gravity. When this was dramatically confirmed in an experiment during a 1919 solar eclipse, Einstein became famous virtually overnight.

General relativity was the summit of Einstein's scientific creativity. It laid the foundations for modern cosmology and extended his earlier mind-bending special relativity theory, which transformed forever our understanding of space, time and motion. In addition, he contributed to the quantum revolution and gave the world the key to the nuclear era with his famous equation, $E = mc^2$. What more could a scientist hope to achieve short of the ultimate theory, wrapping up the Universe in one unified description?

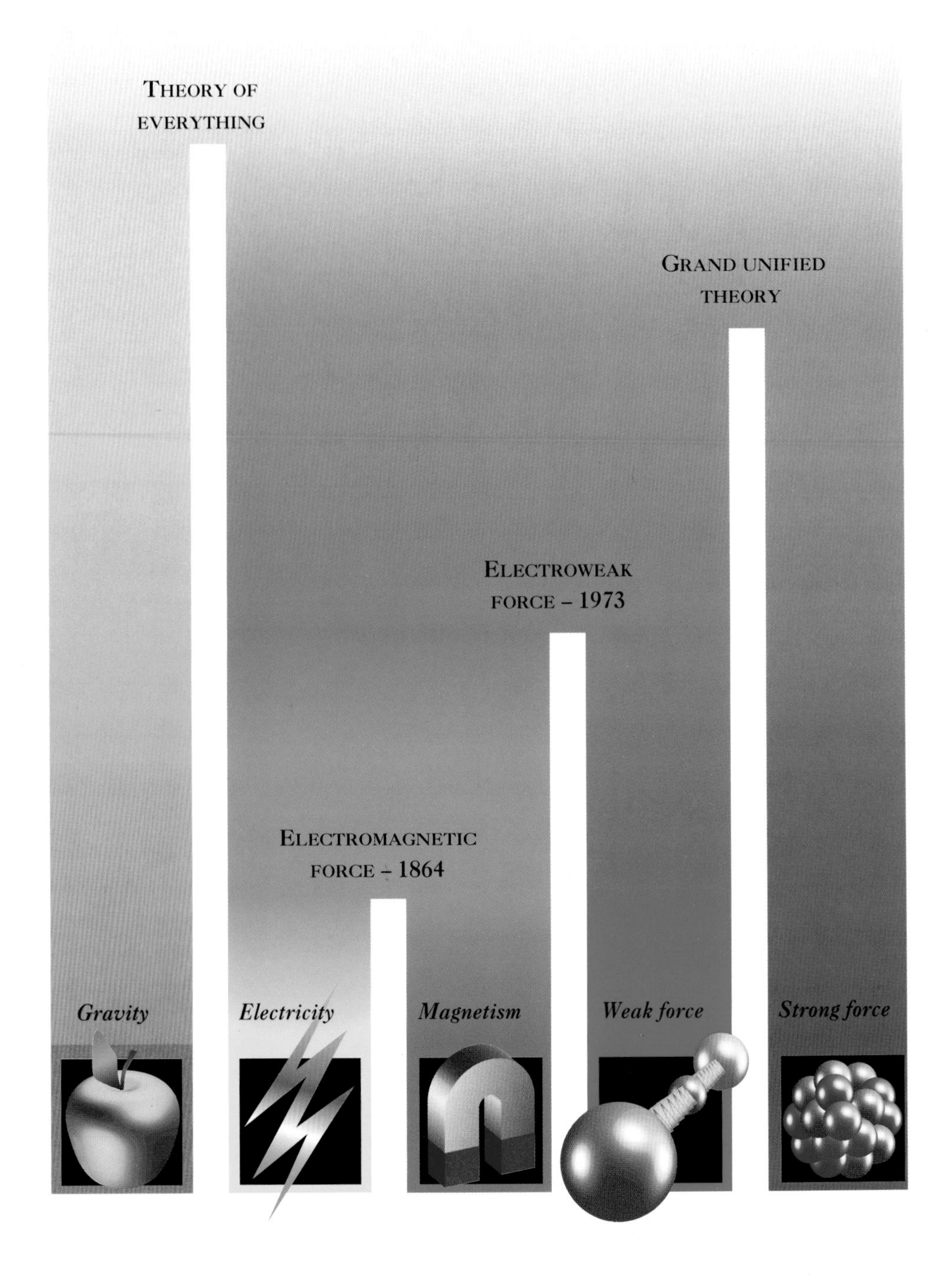

The search for the unification of forces
There are five basic forces of nature. The first unification of forces took place in the 1864 when Maxwell saw that electricity and magnetism are just two expressions of the electromagnetic force responsible for chemical reactions and the transmission of light. The second unification happened in 1973 with the discovery of the electroweak force. The search continues for a single "grand unified theory" that would bring together the electroweak and strong forces, and possibly a "theory of everything", covering gravity as well.

Living legend *With his radical, almost mystical, new ideas that totally transformed our understanding of space and time, Albert Einstein became a legend in his own time, treated with almost god-like reverence. But he remained humble, strangely unaffected by fame, and treated everybody the same way. With a ready sense of humour, Einstein was quick to see the comical side of life. Einstein loved music, especially Bach and Mozart, and played the violin. When asked about Bach by a reporter he answered curtly: "listen, play, love, revere – and keep your mouth shut".*

A FIFTH DIMENSION

In 1921 the German mathematician Theodor Kaluza claimed that gravity and electromagnetism could be unified if Einstein's equations were written in five dimensions, not four. The Swedish physicist Oscar Klein added in 1926 that this extra dimension is in a sense rolled up very tightly and cannot be seen.

Einstein first accepted Kaluza and Klein, then changed his mind. Instead he obstinately followed his own mathematical approach to the unification problem. Initially he was confident that he was on the right track and in 1929 newspapers reported him to be on the verge of discovering the riddle of the Universe. In the excitement he had to hide from the media.

It was a false alarm, and Einstein had to admit he was wrong. In 1931 he told Pauli, who had criticized his theory, "You were right after all, you rascal". Nevertheless Einstein doggedly continued to attack the problem until the end. He died on 18 April 1955, and only the day before, as though preparing for a final effort, had asked for his most recent pages of unification calculations.

THE PHYSICISTS' HOLY GRAIL

When Einstein started working on unification, physics knew only three forces – gravity, electric force and magnetism. The last two had already been unified in the 1860s by Maxwell into electromagnetism. However, in the 1930s two more forces of nature were uncovered: the strong and the weak nuclear force. Any theory describing the whole Universe must include these forces. However, Einstein chose to ignore them because they were described by quantum mechanics, which he detested. For anyone else this would have been scientific suicide, but Einstein continued on. Once asked what criteria he would use on his deathbed to judge his life a success or failure, he replied: "Neither on my deathbed nor before would I ask such a question. Nature is not an engineer or a contractor, and I myself am part of Nature."

Einstein's dream lives on. A new generation of physicists picked up the idea of a unified theory, and 20 years after Einstein's death the next stage in the search for a univeral force picture was achieved: electromagnetism was successfully linked with the weak nuclear force to give the electroweak force.

Building on this electroweak unification success, physicists look towards a "grand unified theory" (GUT), which incorporates the strong nuclear force in the same way. Evidence for such a theory could come from ambitious experiments (see page 84). The ultimate challenge of a "theory of everything" would have to bring gravity into the picture. While all the other forces are quantum-governed, gravity and quantum mechanics have yet to be fully reconciled. Until they are, the idea of a single force remains a dream. The energy scales of these unifications are probably out of reach of laboratory experiments and have to be inferred from lower-energy results or from astrophysics. However, these unifications played an important role in the development of the very early Universe (see page 84).

Symmetry points the way

THE ELECTROWEAK UNIFICATION

In the 1950s, a powerful new way of looking at fundamental forces came from studies of mathematical symmetries. This led to a deeper understanding of how forces work and opened the door to new efforts to unify them. The results had far-reaching implications. Each force has its own symmetry.

Many objects have a regularity that makes them look the same when reflected in a mirror or rotated. The human body has an approximate left–right symmetry, a sphere looks the same from all angles and a snowflake has a hexagonal symmetry.

Such geometrical symmetries are easily recognizable. But symmetry can also be abstract. The laws of nature reflect deep mathematical symmetries underlying the fabric of the Universe. The term "gauge symmetry" applies to certain phenomena that are symmetrical when viewed in the abstract mathematical world.

An example of a gauge symmetry occurs in electromagnetism: an electron travelling between two electrodes at different voltages sees only the voltage difference between electrodes, not the actual voltages. The force on the electron can be viewed from many mathematical angles, but is always the same – that is, "symmetrical". There is no "north pole" of voltage, only relative bearings. Looked at in this way, forces become Nature's way of preserving mathematical symmetry.

In the late 1940s, the deluxe version of electromagnetism arrived. Pioneered by Richard Feynman, Julian Schwinger and Sin-Itiro Tomonaga, and called "quantum electrodynamics" (QED), it incorporates Maxwell's classical electromagnetism with quantum theory and special relativity, the two cornerstones of twentieth-century physics. QED's long-range force messages are transmitted by massless photons flipping back and forth between electric charges (see page 48). The predictions were so accurate – to within a few parts per million – that experimenters had their work cut out to check them.

Six-fold symmetry ***A resin cast of a snowflake crystal. It has a beautiful six-fold symmetry, which comes from a basic pattern (itself left–right symmetric) progressively rotated in 60 degree steps.***

Encouraged by this success, physicists looked for more examples of symmetry at work. In 1956 Julian Schwinger applied gauge symmetry ideas to electromagnetism and the weak force together. Echoing old ideas of Enrico Fermi's, he believed that these two forces were somehow related, and passed this conviction on to his student Sheldon Glashow.

In 1961 Glashow produced a theory combining the two forces. It had three weak-force carrying particles: an electrically charged pair, W^+ and W^-, and a neutral version, Z (or Z^0). They all had to be heavy particles, because the weak force acts at short-range, but their exact masses could not be calculated. Glashow had forced a solution. He had shown how to make a combined "electroweak" theory in principle, but in his own words he had "completely missed the boat" on the problem of explaining the masses of the weak-force carrying particles.

ABDUS SALAM – A WORLD SCIENTIST

At an international conference in Tokyo in 1978, Pakistani physicist Abdus Salam introduced a new physics term: "electroweak" unification. Salam, born in Jhang, – then part of India – in 1926, is at home in many cultures. A practising Muslim, he often quotes from the Koran, which exhorts believers "to study Nature, to reflect, to make the best use of reason and to make the scientific enterprise an integral part of the community's life."

His own early experience made him aware of the problems that result from the lack of resources that are available to scientists from developing countries. Returning to his native Pakistan in 1951 after his initial research career in the United Kingdom, he found himself cut off from the constant stimulation that is necessary for frontier research. In the 1960s, he established the International Centre for Theoretical Physics in Trieste, Italy, which is now a world science rendezvous. The centre provides a valuable foothold for talented young scientists from developing countries at the beginning of their careers.

ELECTROWEAK FORCE

What Glashow did not know at the time was that a symmetric theory can also have non-symmetric outcomes. Building on new concepts from other areas of physics, particle physicists in the mid-1960s saw an alternative route to what is called "spontaneous symmetry breaking". New, unknown heavy particles could make empty space unsymmetric. The weak force carriers "swallow" these heavy particles and acquire mass, while the photon remains massless.

One of the suggestions came from the Scottish physicist Peter Higgs, and ever since the symmetry-breakers have been called "Higgs particles". Seizing on this idea in 1967, Steven Weinberg of Harvard University and Abdus Salam of London's Imperial College independently developed a theory that unified electromagnetism and the weak force.

Initially the idea was not welcome. It looked "unrenormalizable", running into trouble with its mathematics by throwing up infinities to jam the calculations. In the early 1970s, the Dutch theorist Gerard 't Hooft showed how the scheme could be made mathematically sound.

In 1979, Sheldon Glashow, Abdus Salam and Steven Weinberg received the Nobel Prize for Physics for their work in unifying electromagnetism with the weak nuclear force.

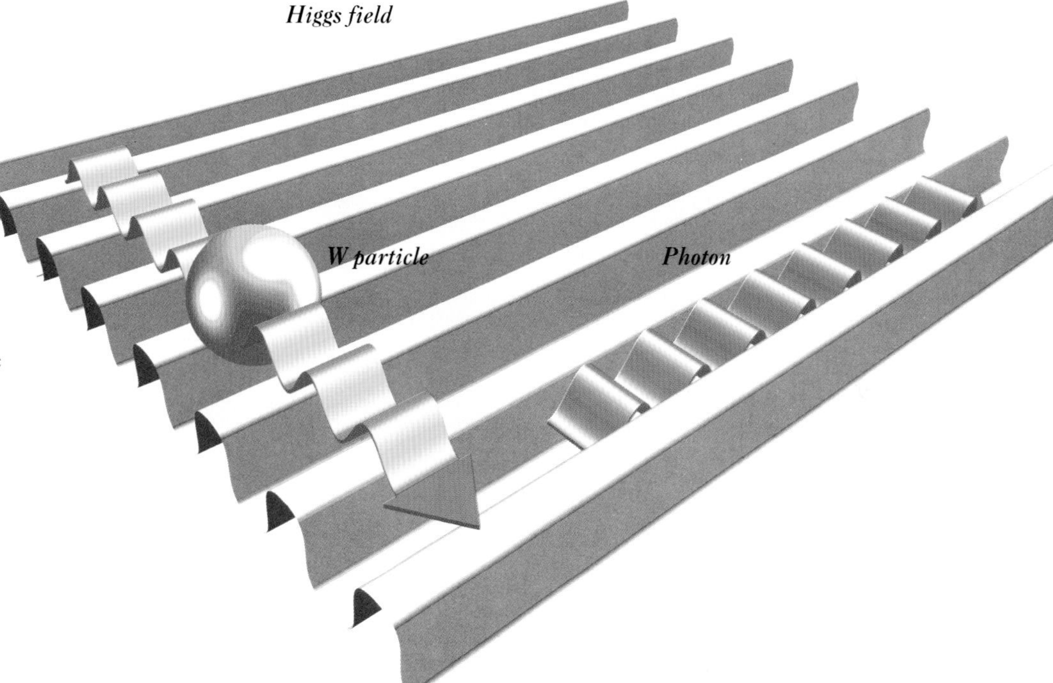

The broken symmetry of space
The electroweak unification comes as a result of the mysterious Higgs particles, which "spontaneously break" the symmetry of empty space. In quantum physics, even a vacuum is not empty; the uncertainty principle says it is full of invisible transient particles, constantly popping on and off. The Higgs particles give empty space a texture – the so-called Higgs field – with properties something like the corrugations hidden inside a sheet of cardboard. The massless carriers of the electromagnetic force (photons) travel along the hidden corrugations and do so easily, but carriers of the weak force (Ws and Zs) have to cross the corrugations and need a lot of extra energy, which they absorb from the Higgs field, so becoming heavy. Without the Higgs field both the electromagnetic and weak force carriers would be massless. While physicists are sure that the Higgs particles exist, what they are and how they operate is a mystery. The discovery of these particles, which eerily permeate the vacuum, is one of the main objectives of high-energy physics today.

Gargamelle's ghost

THE NEUTRAL CURRENT

Until 1973, the weak nuclear force had always been seen to swap electric charges around. Then an experiment at CERN saw something new. The weak force had "jolted" particles without changing any electric charges. Suddenly electroweak unification was in business.

Ever since Fermi's initial explanation in 1934, physicists were convinced that the weak nuclear force, the cause of beta radioactivity, had to be carried by particles that were heavy and electrically charged: heavy, because the force has only a very short range, so the messenger particles have to be correspondingly heavy; and electrically charged, because weak nuclear phenomena apparently always shuffled around electric charges.

The unification of electromagnetism and the weak force, developed in the 1960s by Sheldon Glashow, Steven Weinberg and Abdus Salam, predicted that the heavy, electrically charged carrier of the weak force, called the W particle, should also have an electrically neutral counterpart, the Z. The theory said that Ws and Zs both contributed to weak nuclear processes. The trouble was nobody had seen a weak nuclear exchange that left electric charges intact, a so-called "neutral current". Or nobody thought they had.

In their classic two-neutrino experiment of 1962, Leon Lederman, Jack Steinberger and Mel Schwartz had noticed a few neutrino collisions with no accompanying charged particle tracks. The collisions had at first been dismissed as background "junk" caused by unwanted particles straying out of the massive steel shielding. But perhaps they were examples of neutral current.

Convinced that electroweak unification was on the right path, Steven Weinberg urged experimenters to search harder for neutral currents. The conviction grew when in 1971 Gerard 't Hooft showed that the unification scheme was mathematically sound, allowing Feynman-style calculations.

Meanwhile at CERN, a giant new bubble chamber intended for catching neutrinos was being commissioned. It was named Gargamelle, and contained 18 ton(ne)s of the heavy liquid freon. The search for neutral currents was initially very low on the list of Gargamelle priorities.

The photographs of Gargamelle's neutrino interactions were examined by physicists all over Western Europe, but late in 1972 Helmut Faissner's group, which was working in Aachen, Germany, stumbled across a picture-postcard example of a neutral current interaction. The photograph showed where an

Gluttonous giant *While the first generation of neutrino beams got under way, CERN was preparing for the next round. Under an agreement with the French Atomic Energy Commission, a giant bubble chamber was built, 5 m (16 ft) long and weighing 25 ton(ne)s, to hold 18 ton(ne)s of liquid freon. The huge chamber (seen here inside its yellow-painted electromagnet) was named Gargamelle, after the mother of the gluttonous giant Gargantua in Rabelais's classic book. She gave birth to her titantic offspring through her ear after consuming a huge amount of tripe, followed by a potion which paralysed her sphincter.*

invisible neutrino had passed right through the chamber, but in its wake had severely jolted an electron.

Taking the photograph to a physics meeting in Oxford, Faissner was met at London's Heathrow Airport by Donald Perkins, where the two physicists immediately retired to the bar to examine the photograph and celebrate their success.

CERN management, unused to big discoveries, was not convinced. Other experiments were not seeing neutral currents, and the Gargamelle team were politely asked if they would like to withdraw the result. The principal members were adamant that they were right and, in 1973 after much soul searching, the discovery was announced at an international meeting at Aix-en-Provence, France. Abdus Salam, an architect of the electroweak unification, described the atmosphere as "like a carnival".

Discovery of neutral current, 1973. *Entering the bubble chamber from the right, a neutrino has nudged an electron from an atomic orbit. This electron goes on to jolt others. The round "eyes" are the lamps illuminating the chamber. Neutral current has virtually nothing to do with everyday life. Yet this wispy effect powers the biggest explosions in the Universe, the supernovae. This link between tiny terrestrial clues and major astronomical upheavals is typical of the new understanding of basic physics.*

Weak interactions *Left, in classic beta decay, a neutron (two "down" quarks and one "up" quark) is transformed into a proton – the exchanged W particle converts a neutron down quark into an up quark, while the neutrino becomes an electron. In a neutral current interaction mediated by a Z particle (right), a neutrino jolts a quark or, as shown here, an electron. In high-energy experiments, the two processes operate in parallel. Even before the Z particle had first been seen, comparing the neutral current with electromagnetic effects had enabled physicists to make an estimate of the particle's mass.*

Charm school for quarks

DISCOVERY OF THE CHARM QUARK

The original quark family had just three members. However, to make everything work out right there were soon suggestions of a fourth quark, "charm". Its existence was confirmed in two parallel US experiments in 1974 uncovering an extraordinary new particle.

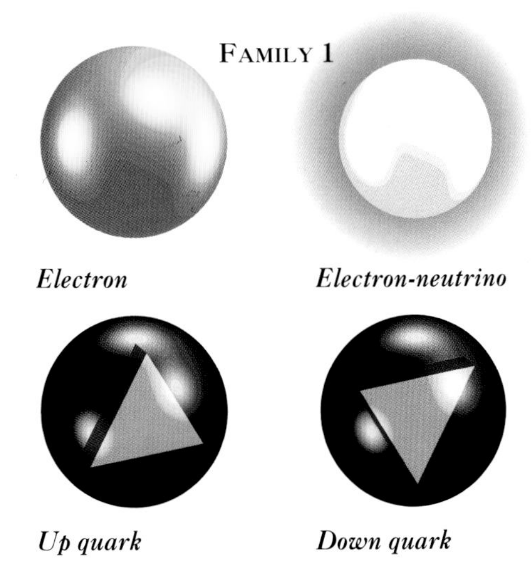

The two groups of fundamental particles known in the 1960s did not match. While there were three quarks (up, down and strange), the lepton family had four members (the electron, the muon and their accompanying neutrinos). This imbalance troubled Sheldon Glashow and James D. Bjorken, who suggested as early as 1964 that there ought to be a fourth quark. They called it charm because it added a pleasing symmetry to the subatomic world.

There was no need at the time for an additional quark to build any known particle. However, in the late 1960s the case for charm was strengthened. The proposed unification of electromagnetism and the weak nuclear force initially involved only the leptons. The quarks were left out. Glashow, working this time with Luciano Maiani and John Iliopoulos, showed how the unification could be extended to quarks, but only if there were four of them. Most physicists thought the idea was far-fetched and there was no organized hunt. In any case, nobody knew where to look.

TING AND TEAM

Professor Samuel Ting epitomizes the dedication needed to do modern physics. Born in 1936, Ting shared, with Burton Richter of Stanford, the 1976 Nobel Physics Prize for their independent 1974 "November Revolution" discoveries of the J/psi particle.

"We were not looking for a new quark when we were doing our 1974 experiment," says Ting. "The story really started in 1966 when an experiment at the Cambridge, Massachusetts, electron accelerator apparently showed a deviation from quantum electrodynamics, suggesting that the electron has a size."

Ting checked this result at the DESY electron accelerator in Hamburg, and found the Cambridge experiment was wrong. The electron had no detectable size.

However, during these measurements he found something interesting. A photon coming out of a high-energy collision sometimes may transform into a heavy particle – a rho, an omega or a phi. Ting continued to study these particles, and after a few years wondered how they had about the mass of a proton. Why not heavier? "This was our motivation for going to higher energies," explains Ting.

"If you look at a town like Geneva when it is raining, there are about ten billion raindrops falling per second. Imagine that one of the raindrops is pink and you have to go and find it. That gives a picture of the problem we were dealing with."

Here, Samuel Ting proudly displays the sharp spike of his particle (initially called "J") discovered at Brookhaven.

Completing the picture *Until 1974 all matter was thought to be composed of just four leptons and three quarks. Theoretical ideas suggested there ought to be a fourth quark to complete the picture. Physicists even had a name for it: the charm quark. It was seen for the first time in the J/psi particle, discovered in 1974, giving a new correspondence between quarks and leptons.*

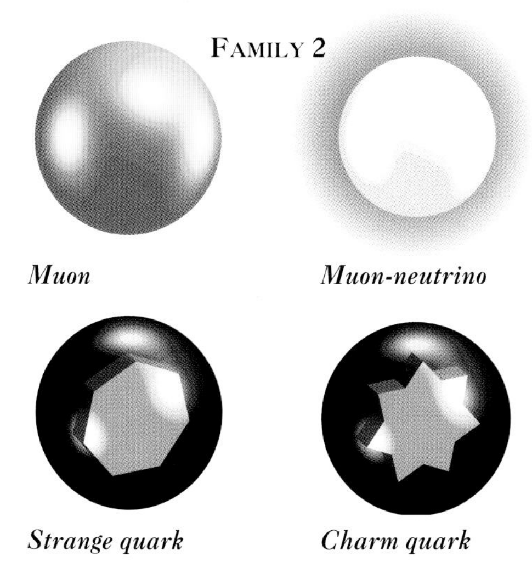

Muon *Muon-neutrino*

Strange quark *Charm quark*

NOVEMBER REVOLUTION

In November 1974 the situation changed dramatically. A group at Stanford led by Burton Richter was running an experiment in SPEAR, a modest new machine fed by the mighty linear accelerator. By slowly varying the energy of SPEAR's colliding electron and positron beams, the experimenters found a sudden increase in their signal at one particular energy – it was a new particle. Richter named it "psi" and announced the discovery at a seminar at Stanford on 11 November.

In the seminar audience was Samuel Ting from MIT. Throughout the summer of 1974 his team, working at Brookhaven, had been collecting evidence that pointed in the same direction. They were studying pairs of muons produced from the collisions of protons and nucleons and had stumbled across a huge peak in a small corner of the recorded data. Stunned by the size of the signal, the team was busy rechecking their results to rule out any possibility of a technical error. However, when Ting arrived at Stanford and learned of Richter's discovery, he realized he had found the same particle. Ting preferred to call it the "J".

Acknowledging the dual discovery, the new particle is universally known as the "J/psi". Its sudden appearance on the scene is commonly referred to as the "November Revolution" of physics.

Explaining the J/psi was less easy. By the time the long arguments had died down, everyone agreed that the fourth quark, charm, had turned up. In the J/psi, the charm quark was clutching its own antiquark. With zero net charm, the new quark label was initially hard to see.

STILL MORE QUARKS!

With four quarks and four leptons, physics now looked very neat. But this happy balance did not last long. In 1975 Martin Perl, also at Stanford, found a new lepton called the tau , a superheavy relative of the electron and the muon. There also had to be a tau-neutrino. Suddenly the number of leptons had jumped to six.

To restore equilibrium needed two extra quarks; these were known either as "top" and "bottom" or as "truth" and "beauty". The bottom quark was found in the summer of 1977. Like charm, it turned up "hidden" with its antiquark inside a very heavy particle, the upsilon, discovered by Leon Lederman and a group of physicists at Fermilab near Chicago.

The impressive consistency of this six-fold picture even predicted the mass of the top quark. The unsolved question is why it has to be so heavy, almost 200 times the mass of the proton and comparable to the the nucleus of a gold atom.

The SPEAR ring *The site of the rival "November Revolution" experiment led by Burton Richter, which discovered the first particle containing a charm quark, the SPEAR ring at Stanford was built on a parking lot. The Stanford team called the particle the "psi", but it was renamed the "J/psi" to commemorate its dual discovery.*

Life imprisonment

THE COLOUR FORCE BETWEEN QUARKS

Quarks are imprisoned for life, permanently confined inside particles by an extremely powerful force arising from their "colour" charge. This force is unlike any other. It gets weaker as the quarks get closer together, and stronger if they become separated.

With the Eightfold Way (see page 60), quarks and the omega minus, physicists had been preoccupied with the strong force in the early 1960s, but the ideas of force fields had then been in disgrace. The success of the unified electroweak force reminded them of the power of field theory, and in the early 1970s theoretical physicists switched their attention back to the strong force. Attacking the problem of what binds quarks inside protons and neutrons, they developed a theory, using the quark's special "colour" charge (see page 60). The theory was called "quantum chromodynamics" (QCD), from "chromos", the Greek for "colour".

QCD is comparable to quantum electrodynamics (QED), the relativistic quantum field theory of electromagnetism, but has a more complicated structure. The inter-quark "colour" force is transmitted by eight carrier particles called "gluons", which stick to the quarks like a kind of super-adhesive. In QCD, charge-like properties also depend on space and time, and the smaller the volume, the feebler the interaction becomes. Conversely, the larger the volume, the stronger the force, making it practically impossible to separate quarks.

The colour force linking quarks is like a thick elastic string, with the quarks attached to the ends. As long as the string remains slack, the quarks move easily, but they become difficult to shift once the string becomes tight. The colour force between quarks underlies the strong nuclear force, but at a nuclear level the aggregate strong force due to meson exchange is more easily identified than the force between quarks.

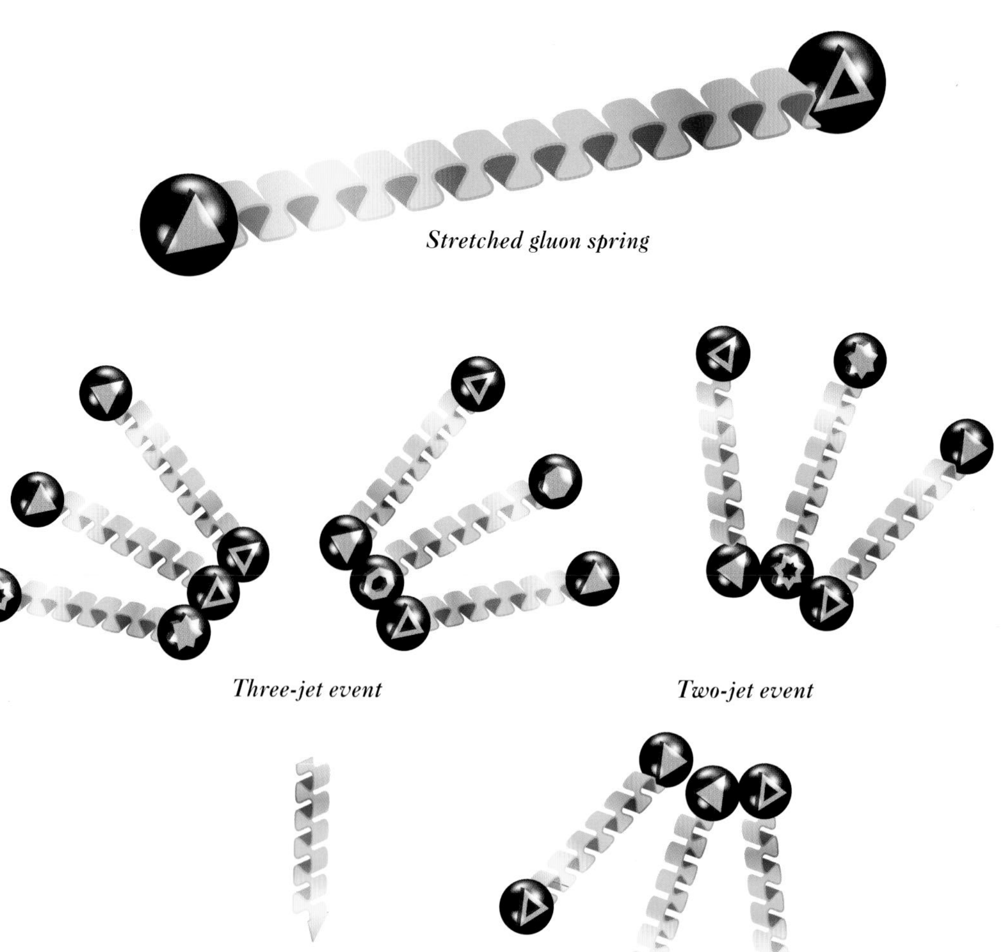

Quark chest expander *The strong bonds between quarks can be likened to a chest expander with very strong springs. When the expander is lying loose on a table, the handles – the quarks – are easy to move. Only when the spring is stretched tightly is any real resistance felt. So it is with quarks: the further apart they are, the stronger the bond is between them. Stretched to the limit by high-energy "excitation", the tortured spring breaks, and the released tension produces two sprays of quarks (right), joined by spring fragments, along the direction of the original stretched spring. Occasionally a piece of gluon spring snaps off (left) producing an additional spray of particles.*

These quark–gluon sprays, known as "jets", soon provided a new probe of the otherwise invisible quarks deep inside the colliding particles. Combinations of jets and other particles gave characteristic "signatures" of new particle interactions.

SAU-LAN WU

A native of Hong Kong, Sau-Lan Wu played a crucial part in the discovery of gluon phenomena in 1979, when a telltale three-jet pattern first showed up in the TASSO experiment at the then-new electron–positron collider, known as PETRA, at the DESY laboratory in Hamburg. Earlier in her career she had worked with Sam Ting on his famous 1974 experiment that played a major part in the discovery of the J/psi and so the charm quark. She is just visible on the right in the photograph of Ting and his team on page 70.

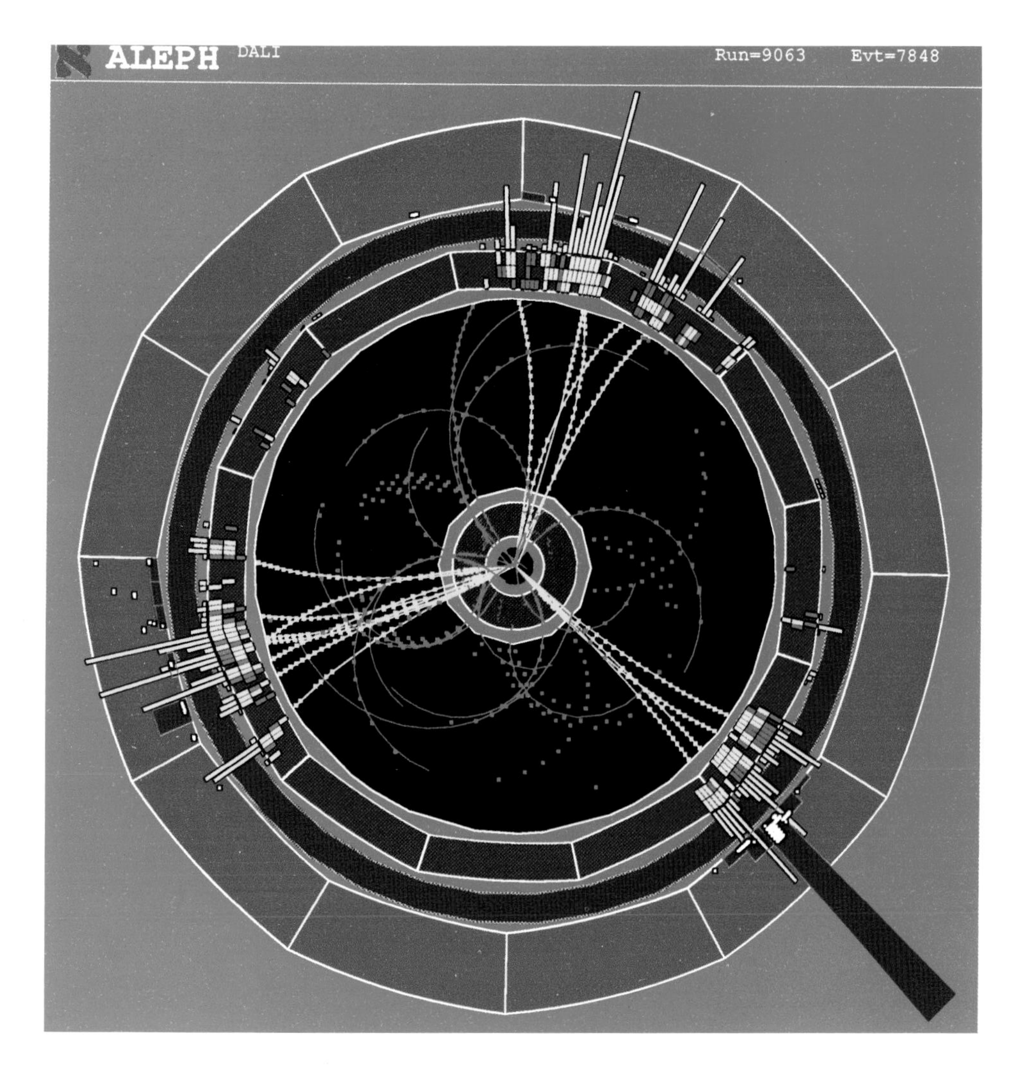

Testing the subnuclear glue *Three clearly visible sprays, or "jets", of particles emerge from this electron–positron collision seen by the ALEPH detector at CERN's LEP electron–positron ring. The different coloured radial zones correspond to different elements of the 12-m (40-ft) diameter detector. However, the inner region (dark background) has been artificially enlarged here to give a "fish-eye" view, which shows better the patterns of the emerging tracks. Further out, the narrow blocks of colour denote the energy released as particles are absorbed in the outer rings of the detector. Two of these jets come from a quark–antiquark system formed from the burst of energy created when the colliding electron and positron annihilate each other. The third jet is due to a gluon, liberated from a snapped-off piece of the overstretched bond linking the quark and antiquark. The gluon jet is probably that on the lower left.*

The colour force is by far the strongest force in nature. It is 1,000 times stronger than electromagnetism, which binds electrons to the atomic nucleus. The energy needed to separate two quarks is equivalent to the energy needed to lift a ton(ne) weight by 1 metre (3 feet). Not even an accelerator as big as Earth's circumference could smash quarks apart.

The paradox of QCD is that this enormously strong force is feeble at the very short inter-quark distances inside the proton and other particles. As long as quarks stay that close, happily bound in triplets or pairs, they are more or less independent particles and enjoy their "asymptotic freedom". If the quarks are hit hard enough, the interaction is so fast that they have no time to stretch apart and feel the full strength of the colour force. Data from experiments like that at Stanford (see page 70) are good fuel for QCD, and in this way physicists are able to do quark calculations without ever seeing quarks.

JET SET

With the colour force so strong, free quarks have never been seen and probably never will be. But when the "elastic" between them is severely tested in a high-energy collision and becomes very tight, the quark bonds "jangle" and eventually snap. When this happens, the energy stored in the elastic is radiated as quarks and antiquarks joined by new bits of elastic. This radiation depends both on the quarks and the intervening elastic, and is seen as narrow "jets" of new particles emerging from the collisions. These jets "remember" the direction in which the elastic was stretched.

A glimpse of creation

SPOTTING THE WEAK FORCE CARRIERS

In 1983 the heavy W and Z particles predicted by the electroweak theory were discovered at CERN. It had called for revolutionary accelerator ideas to collide protons with their antimatter counterparts, antiprotons, recreating the conditions of the Universe when it was only a fraction of a second old.

Although Steven Weinberg, Abdus Salam and Sheldon Glashow had been awarded the 1979 Nobel prize for physics for their electroweak theory, this was a somewhat daring decision by the Nobel committee. The theory could not be put into the textbooks until the W and Z particles, predicted as the carriers of the weak nuclear force, had been found. They were very heavy, roughly 100 times as heavy as a proton. Finding that much energy was going to be difficult.

In the mid-1970s two new "super" proton synchrotrons had come into operation, one at CERN and the other at a new US Laboratory, Fermilab, just outside Chicago. Both were about 7 kilometres (4 miles) in circumference and took protons to over 400 GeV before slamming them into fixed targets. This is an inefficient way of using the energy fed into the protons at great expense, because most is lost in the recoil of the target. As they stood, neither machine had a chance of finding Ws and Zs.

A more efficient way of using particle energy is to fire two beams at each other, and the most cunning way of all is to store two beams circulating in opposite directions, one of particles and the other of the corresponding antiparticles, in the same ring. Using electrons and positrons, this was first done at the Frascati Laboratory in Italy in 1963.

In 1976 Carlo Rubbia, a 42-year-old Italian physicist teaching at Harvard University in the United States, suggested adapting an existing accelerator to collide protons and antiprotons in the same ring. The problem was that nobody had ever collected enough antiprotons for such a scheme to be possible.

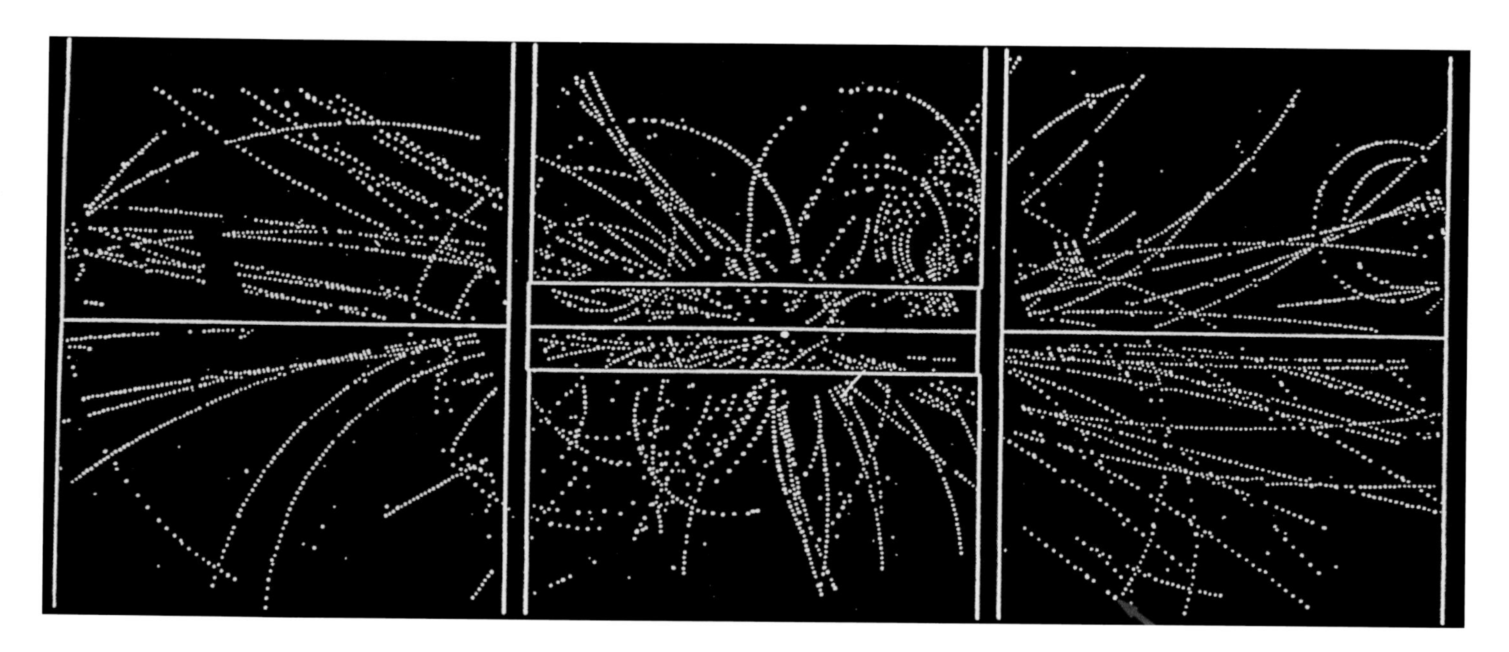

MASTERING ANTIMATTER

Rubbia first took his proposal to Fermilab, but was turned down flat. He then put his idea on the table at CERN, where Simon van der Meer, a quiet but brilliant Dutch accelerator physicist had earlier developed a method for collecting and taming particles called "stochastic cooling". This provided the key to Rubbia's plan.

In 1978 the CERN management, aware that CERN had missed major discoveries by playing safe, boldly gave the go-ahead. CERN's Super Proton Synchrotron (SPS) was converted into a proton–antiproton ring, and the world's first antiproton factory was built.

Two major experiments were prepared, codenamed UA1 and UA2, where the initials "UA" stood for "underground area". Vast underground caverns had to be excavated to house the multi-layered detectors that surrounded each collision point. UA1, the larger experiment, had 2,000 ton(ne)s of concentric boxes, like a high-technology Russian doll. The detector was pieced together by 140 physicists from 12 research centres – 11 in Europe and one in the United States – mercilessly driven by Rubbia.

The collider started running in 1981. By Christmas 1982 the physicists had confident smiles, and the first rumours of success were heard. In January 1983 came the announcement, first from UA1. Out of about 1,000 million proton–antiproton collisions, only about a tenth of one per cent provided the right conditions to create W particles. A couple of months later the even more difficult Z particles were also spotted.

The energy producing the W and Z particles was equivalent to the conditions just a 1,000 millionth of a second after the Universe was born in the Big Bang. The following year, 1984, Rubbia and van der Meer shared the Nobel prize in physics, one of the shortest intervals ever between discovery and the award.

Missing energy (top) *Neutrinos are resolutely invisible, but there are ways of "seeing" them. Neutrinos carry off energy, and physicists compare the energy coming off on opposite sides of a collision; "missing energy" shows that a neutrino has escaped. Here, a picture from UA1 at CERN shows the decay of a W particle giving an electron (arrowed in red) emerging back-to-back with a sliver of "missing energy".*

UA1 detector (left) *The 2,000-ton(ne) detector, here taken apart for routine maintenance, pioneered a new dimension in collaboration for laboratory experiments, with some 140 scientists involved.*

CARLO RUBBIA – LIFE AT 40 MILES PER HOUR

Carlo Rubbia was born in the small town of Gorizia, Italy, in 1934. His father was an electrical engineer at the local telephone company and as a boy Carlo became deeply interested in scientific ideas. Colourful and controversial, he was CERN's Director General from 1989 to 1993. Rubbia describes himself as an internationalist, and he is an incessant air traveller – the airline Alitalia voted him an honorary member of its board. Friends calculated that he jets around so much that his lifetime's average velocity is over 40 miles per hour! But Rubbia's great strength is physics, where his insight and imagination ensure he is permanently one scientific step ahead.

The Standard Model

FAMILIES OF ELEMENTARY PARTICLES

Particle physics is now packaged as the "Standard Model", a handy way of understanding the fundamental processes of the Universe. But with the strong and electroweak forces resolutely independent and containing many unknown quantities, the packaging could be better.

All matter is built from 12 elementary or basic particles. These are the six quarks, which feel the strong nuclear force, and six leptons, which do not. The twelve particles are also divided into three "families" of four, each with two quarks and a pair of closely-related leptons.

The first such family (the up and down quarks, the electron and its accompanying neutrino) makes up ordinary matter and accounts for everyday phenomena. Up and down quarks form protons and neutrons, which cluster into atomic nuclei. Nuclei attract electrons forming atoms and atoms group into molecules. In this way 92 elements and more than half a million chemical compounds are made, resulting in the rich variety of material that makes up the world around us.

The two other families, modelled on the first, make unstable particles that show up

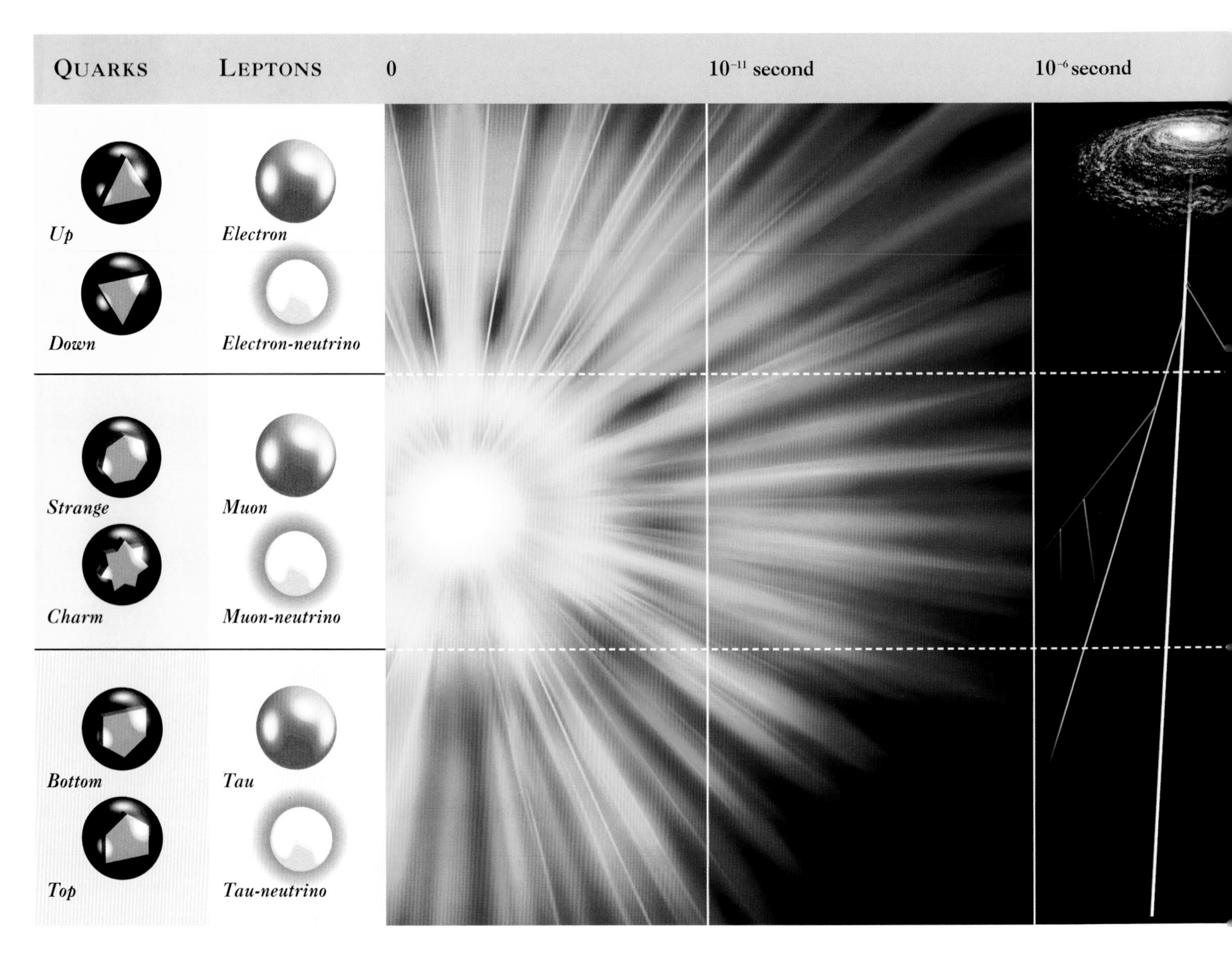

Family resemblance *All matter can be understood in terms of six quarks and six leptons, grouped into three families, each containing two quarks and two related leptons. The first family underlies everyday matter, while the second and third come into play only at high energies – in cosmic rays or laboratory experiments, or in the extreme temperatures of the first few seconds of the Universe after the Big Bang (see page 104).*

Quarks feel all three forces – strong, electromagnetic and weak. Leptons always feel the weak force; if they are electrically charged (the electron, the muon and the tau) they also feel electromagnetism, but neutrinos, which are electrically neutral, feel only the weak force.

MESSENGER PARTICLES

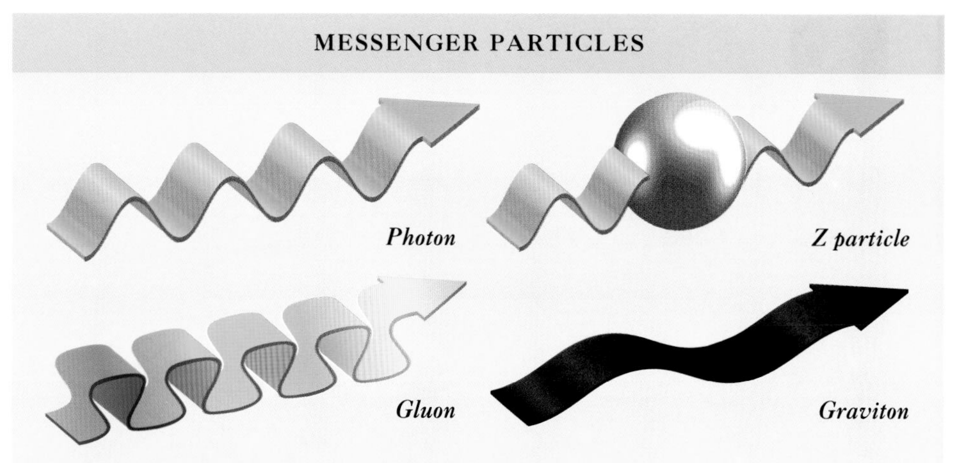

Each force is carried by a particular messenger particle: the photon for electromagnetism, the Ws and the Z for the weak force, the gluon for the strong quark force and the graviton for gravity.

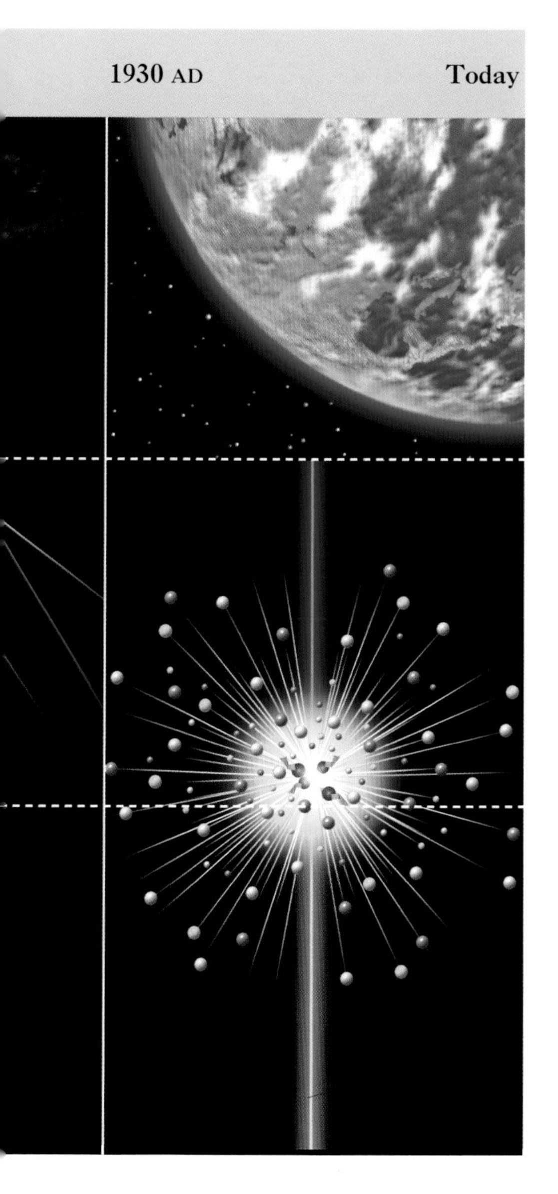

only in cosmic rays and high-energy experiments. The second family groups the strange and charm quarks, the muon and the muon-neutrino, while the third contains the top and bottom quarks, the tau and the tau-neutrino.

This is the static "chassis" of the Standard Model. Its motive power comes from two independent force "engines", with additional particles providing the transmission. One source of power is the electroweak unification of electromagnetism (transmitted by photons) and the weak force (transmitted by W and Z particles). The other is quantum chromodynamics, for the strong inter-quark forces, transmitted by gluons.

The resulting predictions are in line with all evidence from experiments, and powerful cross-checks give precision limits on quantities still difficult to measure. However, the Standard Model cannot be the end of the story; there are too many loose ends. It does not include gravity, the force dominating the large-scale Universe, and it has 20 free parameters – properties and values that it is not able to pin down and which can only be fixed by measurement. These include the number and masses of its particles and the strengths of the forces. In principle, the Standard Model could accommodate three or more quark-lepton families, but in 1979 experiments at CERN's LEP electron–positron collider concluded that there are just three quark–lepton families (see page 78). The electroweak and strong forces coexist in the Standard Model, but are not linked in any way. These two facets of the Standard Model would become unified in a wider grand unified theory (see page 64), while incorporating gravity would have to await the ultimate theory of everything.

THE MISSING HIGGS

Particle masses come from the Higgs mechanism, which affects everything, even the vacuum, subtly breaking the underlying electroweak symmetry (see page 66). Its messenger (Higgs) particles are swallowed by the electroweak particles, which become heavy.

The paradox of the Standard Model is that while everything else comes out right, the Higgs mechanism remains a complete mystery. The missing ingredient in today's physics, it is the main goal of all new high energy projects. Opinions differ about the Higgs particles. Some physicists say they could be made up from known electroweak particles; others think they are new elementary units. Whatever the Higgs mechanism is, it could be quite different to anything seen so far. Seeing the Higgs is the challenge for new "supermachines" for the twenty-first century (see page 86).

The Z factory

THE LEP SYNCHROTRON

On 14 July 1989, the 200th anniversary of the French revolution, CERN was the scene of a revolution of a very different kind as the first particles went round LEP, the Large Electron Positron collider. This 27-kilometre (17-mile) ring is the world's biggest particle accelerator.

At the foot of the Jura mountains north-west of Geneva lies the world's largest particle accelerator. However, despite its size, CERN's 27-kilometre (17-mile) LEP (Large Electron Positron collider) ring is largely invisible in its 3.8-metre (12-foot) wide tunnel about 100 metres (330 feet) below the rural landscape of the Franco–Swiss border. Beams of electrons and positrons, born in the CERN main site on the Swiss side, pass through an interlinked chain of accelerators before being injected into LEP in France, and circle the ring in opposite directions at almost the speed of light.

The counter-rotating particles collide at four points. There the narrow tunnel opens out into cathedral-sized caverns, the homes of LEP's "eyes", or detectors, huge multi-layered structures, each as big as a house, with almost mystical names: ALEPH, DELPHI, L3, and OPAL. The particles meet head-on at a combined energy of 100,000 million electron volts (100 GeV), annihilating into bursts of energy that materialize as showers of new particles, recorded by the waiting detectors.

Plans were put together for a giant new ring at CERN in the mid-1970s to continue the laboratory's tradition of interlinked circular accelerators, each one feeding the next. CERN had never accelerated electrons and positrons, and storing them at high energy in a ring posed a special challenge. Because these particles are so light, they "skid" much more than protons, losing energy as they go round bends by emitting so-called synchrotron radiation and becoming difficult to accelerate. To overcome this, LEP's ring had to be large.

Aerial view *Hemmed in by Geneva airport on one side and the Jura mountains on the other, the countryside around CERN is far from flat. But by making the ring as compact as possible and by drilling its underground tunnel on a slant, it was just possible to build LEP without having to go too deep under the Jura mountains, where the tunnel is 180 m (590 ft) below ground. Also between Geneva airport and the Jura mountains is the Swiss–French frontier. The CERN site and the LEP ring symbolically straddle this line. The large ring is LEP, 27 km (17 miles) in circumference. Within it is the much smaller SPS ring, 7 km (4 miles) in circumference. As well as acting as an injector for LEP, the SPS also handles protons and other kinds of beam.*

HIGH TIDE AT LEP

For precision measurements at LEP, the energy of the circulating beam has to be known as accurately as possible. The circumference of the LEP ring varies by a about a millimetre (a twenty-fifth of an inch) twice in each 24 hours as the Moon's gravity alternately pulls and pushes on the Earth's crust. These "tides" are amplified by the accelerator, changing the energy of the circulating beam by 20 parts per million. This tiny effect is now routinely allowed for.

Major project *LEP was the largest civil engineering project of its day in Europe. The construction work took six years, starting in 1983, and involved the removal of 1.4 million cubic metres (50 million cu ft) of rock and soil, equivalent to about one-third of the Great Pyramid. Three tunnelling machines drove through the rock at about 25 m (80 ft) per day. There was no question of avoiding unforeseen obstacles. The ring as traced on paper had to be faithfully cut through rock and stone and led through unexpected underground water sources. The drilling machines were guided to a precision of 1 centimetre (0.5 inch), and 60,000 ton(ne)s of equipment was installed in the tunnel. LEP has its own underground monorail. The total cost was 1,300 million Swiss francs.*

Although LEP is the world's largest particle accelerator, it is not the most powerful. Proton machines only a fraction the size can take protons to more than 500 GeV, ten times higher than LEP's initial energies. This is because the light electrons in LEP are difficult to hold on a tight course.

One of the major goals was to probe the Standard Model. Were there only three quark-lepton families, or could there be more? LEP would mass-produce Z particles, and by measuring how long the Z lived it would be possible to fix the number of particle families. Additional families would contain quarks and leptons, and these would be too heavy to be detected directly in Z decays, but their accompanying neutrinos would leave their mark.

A LIFE OF UNCERTAINTY

If there are more neutrinos than the three known from the Standard Model, they would provide additional ways for the Z to decay. A decaying particle is like a leaky bucket: the more holes, the faster the water runs out. Measuring the Z lifetime thus provides an indirect way of measuring the number of particle families.

The first LEP collisions took place in August 1989, and soon the accelerator had produced 10,000 Zs. In November of that year the four detector teams had enough evidence to conclude that there are only three particle families – the chance of a fourth family being only one in a thousand. A few months earlier, a team at Stanford's linear accelerator had also arrived at a three-family limit, but with fewer Zs their result still had a 5 per cent chance of being wrong.

LEP now produces millions of Z particles each year, and the three family picture has been established beyond all doubt. New equipment will double the energy of LEP's electron and positron beams so that it can also produce pairs of W particles, the electrically charged counterparts of the Z, and continue its precision study of the Standard Model.

Taking the lid off LEP

THE LARGE ELECTRON POSITRON COLLIDER

In a high vacuum like that of outer space, electrons and positrons race round LEP's 27-kilometre (17-mile) tube at close to the speed of light. Each time they meet, the electrons and positrons run the risk of annihilation, disappearing in a burst of energy inside the mighty detectors.

LEP's electrons are made in basically the same way as the electron beam inside a television tube, with a heated filament giving off a stream of particles. LEP's electrons get their initial kick from a 100-metre (330-foot) linear accelerator, which takes them to an energy of 200 million electron volts (200 MeV).

Getting positrons is harder. First, electrons are fired into a heavy target, generating a burst of gamma-ray photons which in turn "convert" into electron–positron pairs. The positrons are selected by magnetic fields and stored in a ring until enough are available.

The electrons and positrons are collected in carefully separated bunches, each containing many thousands of millions of particles. The bunches wind their way through the two older synchrotrons (PS and SPS), where they are further accelerated to 22,000 million electron volts (22 GeV), then injected into LEP, the electrons going one way and the positrons the other.

ALMOST PERFECT VACUUM

The electrons and positrons circle LEP inside a lead-coated and water-cooled aluminium tube, 10 centimetres (4 inches) across. This contains an almost perfect vacuum, maintained by a coated "getter" strip on the inside of the tube which acts like flypaper on residual air molecules. The vacuum has to be good to avoid collisions with stray molecules. LEP is the longest high-vacuum system in the world – inside it an electron can travel a light-year before bumping into a residual air molecule.

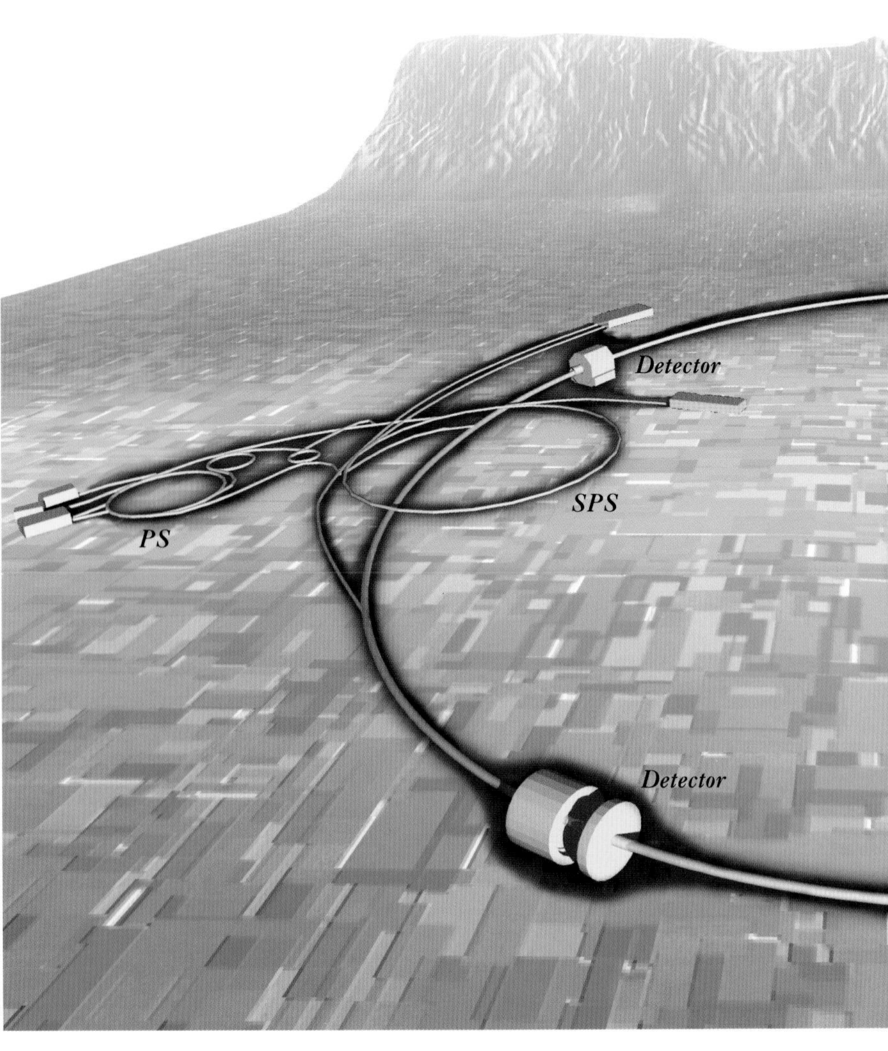

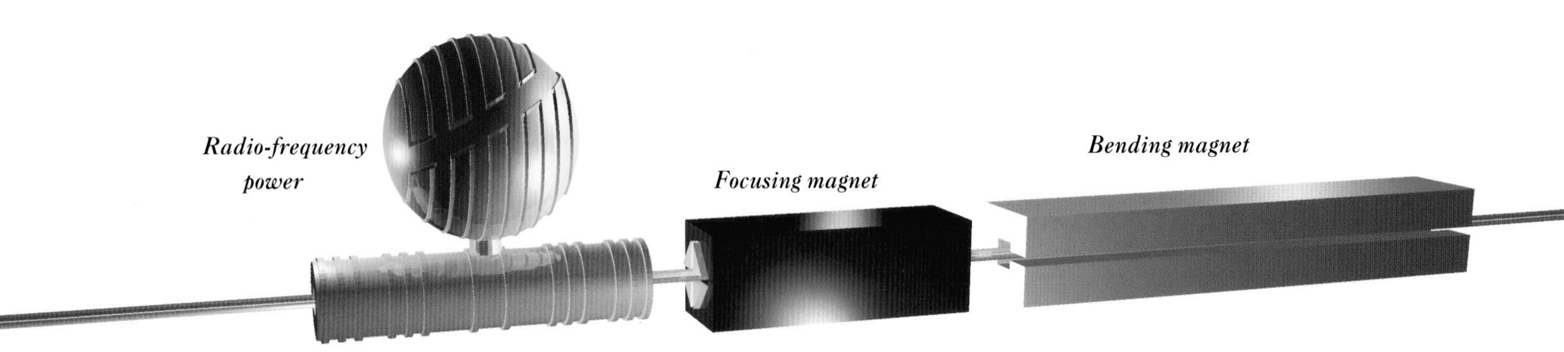

Particles from the vacuum *When LEP's electrons and positrons collide, they annihilate each other and disappear, producing a burst of energy in a vacuum – a space with no visible particles. However, in a vacuum, particles are continually flickering on and off through energy borrowed by the uncertainty principle. If enough energy is supplied, these vacuum denizens can become real particles. In this way LEP manufactures its Z particles. By studying the way Z particles decay, physicists have improved their understanding of how quarks and leptons coexist.*

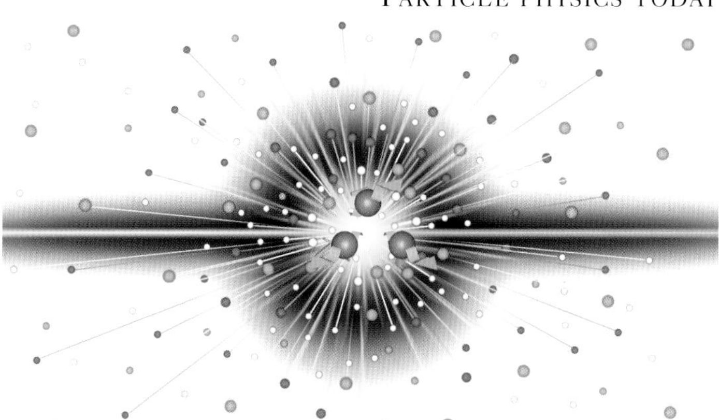

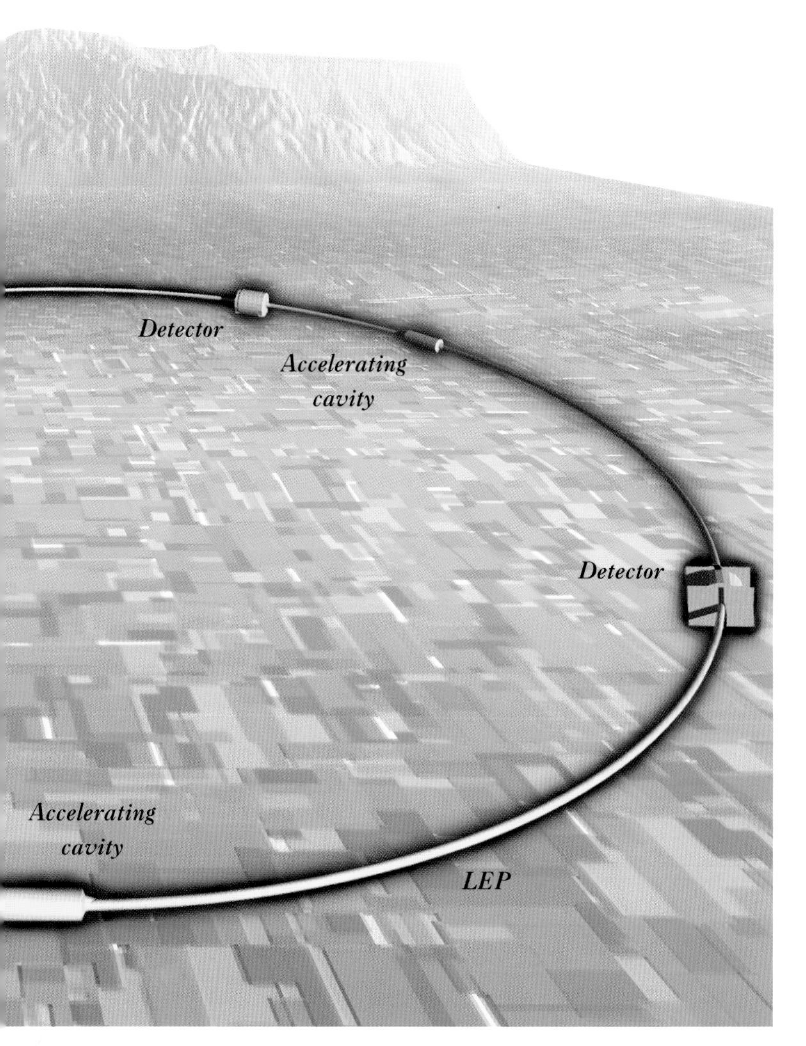

Travelling close to the speed of light, LEP's electrons and positrons make 10,000 revolutions of the 27-kilometre (17-mile) ring each second. While the beams are stored, usually for the best part of a day, they travel a total of 13,000 million kilometres (8,000 million miles), more than twice the distance from Earth to Neptune, a trip which took the *Voyager 2* spacecraft 12 years.

LEP's particles are guided and focused by more than 5,000 magnets, and boosted by radio-frequency power to compensate for the energy lost when turning (synchrotron radiation) and to push the energy still higher.

COLLISION COURSE

During acceleration the electrons and positrons are held on different sides of the high vacuum "road" by electric fields. As they reach the designated collision energy, the fields are switched off, and the counter-rotating particles finally come face to face at the four collision points, each at the heart of a giant detector.

LEP's particle bunches are thin flat slabs – a few centimetres (inches) across, but only a fraction of a centimetre wide – each containing many thousands of millions of electrons or positrons. As they approach the collision point, the bunches are squeezed even flatter by strong magnets. This increases the particle density and improves the chances of collisions. But the particles are so small that even in these compressed bunches, meeting head-on 40,000 times a second, most of the particles fly past each other. Each second produces only a few collisions, and each detector collects a few Z particles per minute.

The stored beams gradually deteriorate, and once or twice a day a decision is taken to "dump" the circulating electrons and positrons, and the ring is filled with fresh particles.

DETECTORS – GIANT MICROSCOPES

LEP's electrons and positrons collide deep inside the machine's four giant cylindrical detectors, each about 10 metres (30 feet) across and 10 metres long, and weighing several thousand ton(ne)s. Although the detailed designs of these giant microscopes are different, they all have concentric layers wrapped round the slim LEP vacuum pipe. The innermost layer – the "vertex detector" – sees the decay of short-lived particles which hardly have time to leave the vacuum chamber before disintegrating. The next layer is the main tracking device, recording the trails left by the particles. Several layers of "calorimeters" then measure the energy carried off by particles, and finally an outer cap intercepts muons. Somewhere in the sandwich is a powerful electromagnet to bend the tracks of charged particles.

The art of seeing the invisible

PARTICLE DETECTORS

Particle collisions can make beautiful pictures, looking like surrealist works of art. Interpreting what happens when subatomic particles smash into each other involves sophisticated electronics and computer technology. However, the final judgement relies on the human eye.

Rutherford used to count particle collisions as flashes on a screen. The next step was to record the collisions on photographic film. The particles left trails as they passed through a gas or a liquid, or generated electric sparks, which were photographed. With experiments at the new accelerators producing millions of pictures, which had to be manually measured and analysed, there was soon a data log jam. Physicists turned increasingly to electronic "eyes" to record more selectively what was going on in their experiments, with computers analysing the data.

Big modern detectors, such those at LEP, have 500,000 electronics channels recording signals from "events", as physicists call particle collisions. As a particle bunch passes through a detector, the electronic logic has just 25-millionths of a second before the next bunch arrives in which to decide whether something interesting has happened. If not, the electronics must be cleared and reset, ready for the next bunch. If the data are promising, the next bunch crossings are ignored while the detector digests its information for a few hundred millionths of a second. Then comes data "readout", all channels are emptied, and all the information on the collision – enough to fill a telephone directory – is transferred to the experiment's computer.

The information can now be analysed at leisure, and the complex track patterns admired on a display. However, with physicists always eager for new results, the LEP experimenters have express analysis lines for rapid identification of the special particle "fingerprints" of interesting and unusual patterns.

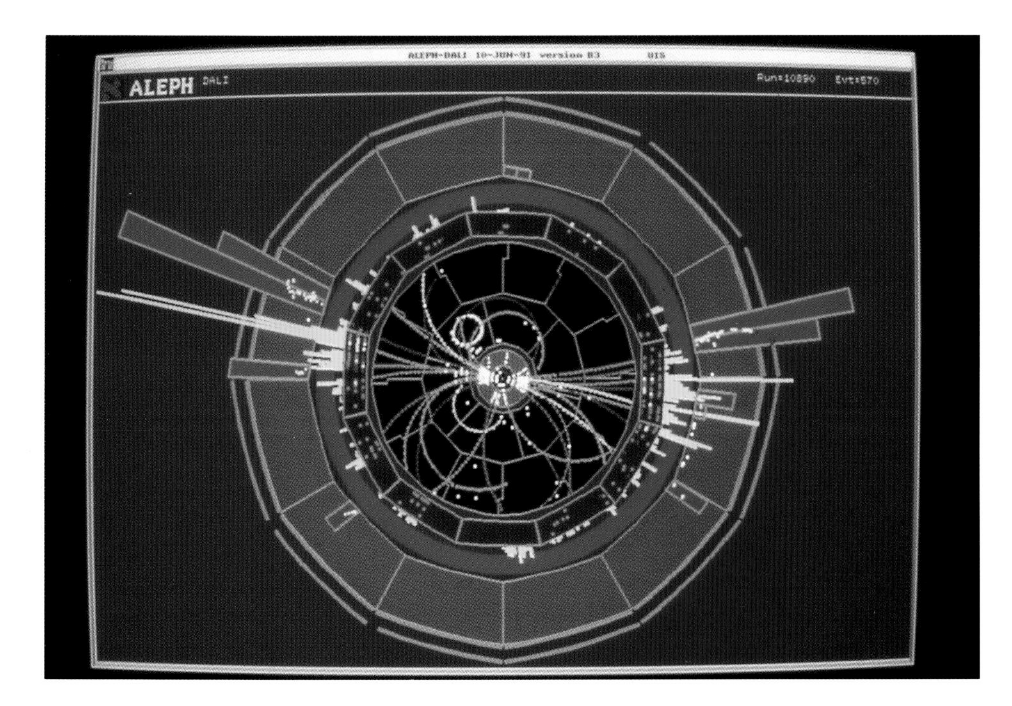

Computer image (above) *A reconstruction of a Z decay in the ALEPH detector at LEP shows the various decay products fanning out from the collision point where the Z is produced at the centre of the detector. The wedge-shaped slices represent the amount of energy carried off.*

End-on view of the ALEPH detector (left) *The central "stained-glass window" is the central tracking chamber where the paths of the emerging particles are picked up. On the outside are the large "readout towers" recording the energies of hadrons.*

DETECTIVE WORK

When a collision is intercepted, the detector registers a cloud of points from the resulting debris, mixed with spurious points from electronic noise. Interpreting this complicated pattern calls for good detective work. The hardest part is to join the points produced by one particle and reconstruct its path or "track" through the detector. Special pattern-recognition computer programs translate familiar point groupings into smooth trajectories.

Other input comes from special detector elements, which identify the new particles produced and analyse them, providing timing, or energy measurements. Inside a magnetic field, a curved track indicates a charged particle, and the curvature shows whether the particle is positive or negative, and enables its momentum to be calculated.

Gradually lines are drawn in this maze as the computer recognizes points that can be grouped into tracks. It is like the child's game of joining dots to reveal a picture – join the wrong dots and the result is unrecognizable. The tracks lead back to the collision, tracing the life history of a particle – maybe a Z.

However, the computer can often only give a rough and ready reconstruction of an event, or get lost in the possible track combinations. An electron can make a tight spiral and the computer can have a hard job picking this out from hundreds of smooth lines. Unresolved results are displayed on a screen for the physicists to interpret, and new patterns fed back to the computer so that it "learns" how to deal with them the next time.

No matter how sophisticated the computer analysis, physicists' experience and judgement remains crucial. Just as Rutherford's trained eye could tell the difference between different sorts of flashes on a phosphorescent screen, so the LEP physicists have the final say when it comes to understanding their electron–positron annihilations.

GEORGE CHARPAK – DETECTOR VIRTUOSO

When Georges Charpak began his career in particle physics, he says, "Physicists were like big game hunters, stalking through the undergrowth hoping to find some rare new specimen. I chose instead to be their 'arms dealer', supplying them with the necessary weapons." Charpak was awarded the 1992 Nobel Physics Prize for his invention of the multi-wire chamber, a totally electronic detector that revolutionized particle physics and now forms the basis of all modern detectors.

Born in Poland in 1924, but a French citizen since 1946, Charpak has a long list of ingenious new techniques to his credit. With bubble chambers, millions of photographs had to be visually examined to select interesting physics. With Charpak's invention, physics experiments became discriminating. For the first time, detectors could be "triggered", ignoring the uninteresting and reacting only to special conditions.

One Charpak ambition is to develop electronic detectors for medical radiography. Able to assimilate information much faster than conventional media, such as photographic film, these techniques can give results more rapidly, with less irradiation for the patient.

Grand unification

PROTON DECAY AND SUPERSYMMETRY

Building on the success of the electroweak theory, in the 1970s physicists tried to extend the unification of forces to bring in the strong force too. Their "grand unified theories" (GUTs) forged a new link between particle physics and cosmology, the study of the origin of the universe.

Following the electroweak unification of electromagnetism and the weak force, the strong force could be brought in by doing the same thing again. The electroweak and strong forces were originally a single "grand unified" force, with an underlying symmetry again broken by a Higgs mechanism. However, this time the energies involved are so big that there is little hope of testing the theory in high-energy experiments.

The simplest grand unified theory, devised by Howard Georgi and Sheldon Glashow in 1974, requires extremely heavy force carriers, called "X particles". These are so heavy, at about 20 millionths of a gram (a millionth of an ounce), that they could be weighed on a sensitive balance! No accelerator can supply the energy to make them. The only place such energies were available was the early Universe, where the grand unified force briefly reigned, just 10^{-34} second after creation.

However, there is an indirect way of testing for GUTs. As well as merging three forces, this unification also links the two classes of elementary particles, quarks and leptons. At the GUT level, quarks and leptons are interchangeable, so protons can decay into leptons. This is a dramatic proposition: the particle forming the very bedrock of our Universe is unstable!

Fortunately the proton's lifetime is at least 10^{32} years, billions of times the age of the Universe, so there is no immediate risk that the world around us will disintegrate. But proton decay is a random process. If enough protons are brought together, there is a slight chance of seeing a proton dying in a characteristic burst of gamma rays. With protons enduring for 10^{32} years, somebody could live for 100 years before there was a good chance that just one proton in his or her body would disappear.

Several experiments to detect proton decay have been launched, and so far none has been successful. But this does not spell the end for GUT theories. The scale of the experiments can be increased to match a longer proton lifetime of at least 10^{35} years. These experiments have to be buried deep underground to screen off cosmic rays that would mask any sign of proton decay.

Despite the appeal of the new GUT theories, physicists were worried about the GUT force carriers, the X particles, being so much heavier than anything else. The W and Z carriers of the weak force are heavy enough, but it was worrying to think of a physics "desert" all the way from 100 GeV to 10^{15} GeV. Surely something must happen in between?

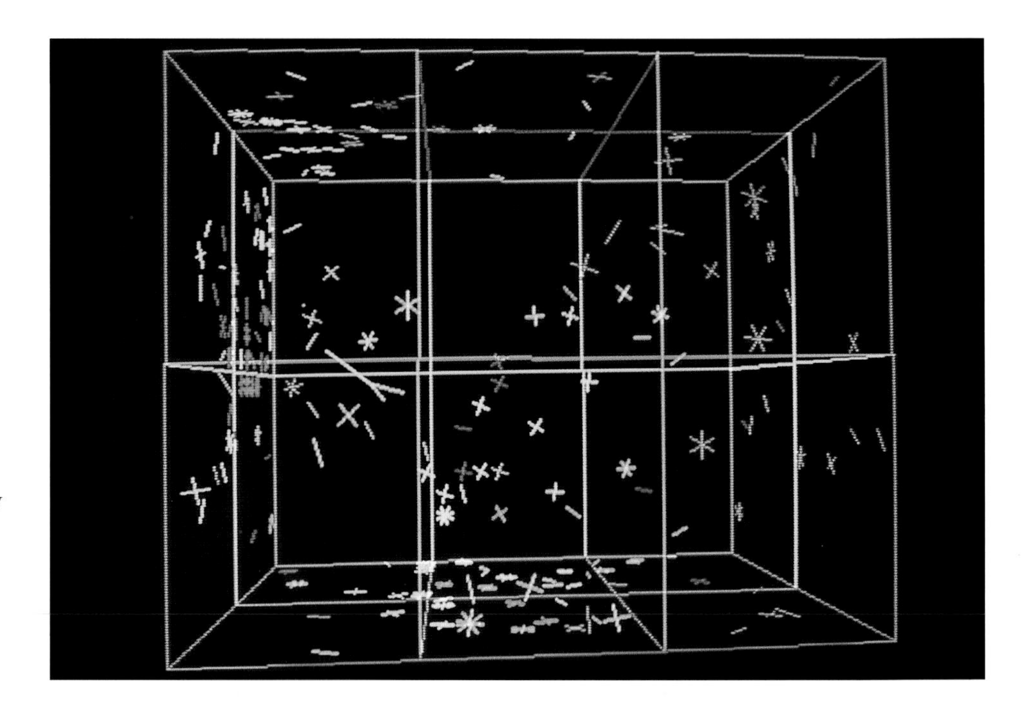

The rarest event in the Universe (above) ***An electronic display of a simulated proton decay in an IBM detector. It simulates a possible simple form of proton decay in which a proton decays into a positron and a neutral pion, which move in opposite directions. Such an event would cause two rings of light to be seen on the walls of the detector. The tracks of the decay particles are reconstructed in the left-hand part of the cube – a positron (short yellow track) and a pion, which immediately decays.***

Deep end (above right) ***A 23-m (75-ft) deep "swimming pool", 600 m (2,000 ft) below Lake Erie in the Morton Salt Mine, Ohio, and filled with 7,000 ton(ne)s of water, has painstakingly searched for proton decay for more than ten years. However, proton decay is probably so slow that even this big detector is too small to see anything.***

The proposal was "supersymmetry", or SUSY for short, linking matter particles (quarks and leptons) and the force-carriers (photons, Ws, Zs and gluons). This scheme creates a new world of super-particles, or "sparticles". Quarks and leptons have super-partners that are known as "squarks" and "sleptons", while photons, Ws, Zs and gluons are twinned by photinos, winos, zinos and gluinos!

Supersymmetric GUTs were the next step, and the ideas could be extended to bring in gravity as well. With this super-gravity picture, physics had its first-ever recipe for handling all the forces in nature. In his inaugural lecture on becoming Lucasian Professor of Mathematics at Cambridge University in 1979, the chair originally occupied by Newton, Stephen Hawking spoke of the end of theoretical physics being in sight.

ANDREI SAKHAROV – MISSING ANTIMATTER

GUT equations suggest a balance between matter and antimatter in the Universe. Why, then, is the Universe apparently composed of matter? Where has all the antimatter gone? An answer was given by the Soviet physicist Andrei Dimitrievitch Sakharov in tne 1960s, before he was exiled to Gorki and his subsequent lamentably brief reappearance on the world stage. Sakharov showed that the Big Bang initially produced equal amounts of matter and antimatter. But a quark asymmetry defined an "arrow of time", making antiparticles more unstable than particles, so that matter quickly gained the upper hand. This effect, which shaped the Universe, has now shrunk to an obscure detail, only detectable through measure-ments of the decays of electrically neutral kaons. To make such an effect possible needed at least six kinds of quark.

Supercooled supercolliders

THE NEXT GENERATION OF ACCELERATORS

In the 1980s, a new generation of proton accelerators appeared that used superconducting electromagnets to provide strong magnetic fields to guide higher energy protons, while at the same time economizing on energy. Physicists had to master a new technology: cryogenics.

In their quest for continually higher energy beams to probe deeper inside the proton, physicists came up against a problem. In contrast to an electron beam, which is light and "flexible", a beam of the much heavier protons is "stiff", and needs to have strong magnetic fields to hold it on course round an accelerator ring. Powering such electromagnets calls for a lot of electricity – a large synchrotron needs about 60 megawatts of current, enough to supply a town of 150,000 people. So physicists turned to a new "green" technology, which greatly reduced power consumption.

When cooled to the temperature of liquid helium – just a few degrees above absolute zero – some metals, such as lead and tin, lose their electrical resistance and become "superconducting". As long as the low temperature lasts, an electric current can flow in these conductors without being replenished. However, superconductivity is delicate because it is easily upset by high currents and high fields, as well as by increases in temperature.

Superconducting electromagnets require far less power, but call for sophisticated cryogenic engineering. The first big superconducting proton accelerator was the 6.4-kilometre (4-mile) Tevatron ring at Fermi National Laboratory, on the Illinois

Wilson's Vision *The Fermi National Accelerator Laboratory near Chicago bears the unmistakable imprint of its founder, the multi-talented Robert Rathbun Wilson. The design of the Laboratory's striking central high-rise building, designed by Wilson and now named after him, is based on the 13th-century cathedral in Beauvais, France, with twin towers and a chancel on the main floor.*

plain just outside Chicago. The Tevatron came into action in 1983 and routinely runs at 900 GeV, using only a third of the electrical power needed to run a ring of conventional magnets at this energy.

In 1992, a new kind of collider came into operation at the DESY Laboratory in Hamburg, Germany. Instead of colliding beams of similar particles, DESY's 6.3-kilometre (4-mile) HERA rings (named after the sister and consort of the Greek god Zeus) collide electrons with protons. The electrons, being extremely small, probe the parts of protons that other beams, with their complicated quark structure, cannot reach. The 820 GeV protons are held in a ring of superconducting magnets, while the 30 GeV electrons have their own separate ring. Cooling the 13 ton(ne)s of liquid helium for HERA's proton magnets called for what was then Europe's biggest liquid helium plant.

21ST-CENTURY SUPERCOLLIDER

Looking towards grand unification, physicists must smash together the quarks and leptons of the Standard Model at still higher energies. Although physicists are not sure what they will find, the symmetry of the equations suggests that 1 TeV (1,000 GeV) is a good goal to aim for.

The HERA electron–proton collider
At the DESY Laboratory in Hamburg, the 6.3 km (4-mile) ring of superconducting magnets takes protons to almost 1,000 GeV. The more compact ring of conventional magnets underneath takes HERA's electrons to some 30 GeV.

With quarks stubbornly locked inside protons, arranging collisions between free quarks is impossible. Each prisoner quark shares the total energy of the proton, so pushing the proton energy as high as possible increases the chances of having 1 TeV quarks inside. A new project is being prepared at CERN for a ring of superconducting magnets in the LEP tunnel (designed from the outset to house two accelerator rings, one on top of the other). This new Large Hadron Collider (LHC) will collide 7 TeV beams of protons, providing the conditions for the tiny quarks hidden deep inside protons to clearly "see" each other.

In contrast to permanently imprisoned quarks, free leptons, particularly electrons, are easy to come by. But producing 1 TeV electrons brings other obstacles. Because the electrons are so light, attempts to accelerate them in conventional circular machines waste a lot of energy in the form of synchrotron radiation (see page 78). Producing 1 TeV electrons would need a ring 1,000 kilometres (600 miles) across!

So, in the hunt for 1 TeV electron beams, physicists have abandoned the circular synchrotron, in favour of the straight or "linear" machine. Two such mighty electron cannons, each several kilometres (miles) long, would fire their beams at each other. With each beam only a hundredth of a hairsbreadth across, this needs precision targeting. Vibrations from tiny earth tremors, traffic or even footsteps would spoil the aim. To keep the two micro-beams locked onto each other the cannon will have to be mounted on automatic jacks, which continually monitor and compensate for micro-seismic effects.

But first new techniques have to be developed to provide the necessary accelerating power and intensity. Preparing and building such huge machines takes a long time. While planning started about a decade ago, these electron and quark supercolliders will come into action only after the year 2000, propelling physics into the twenty-first century. Many of the scientists who will participate in this research are still at school.

Everything with strings attached?

SUPERSTRING THEORIES

What physicists refer to as "particles" might not be particles at all. Theory suggests they behave more like extended objects which can be pictured as pieces of string. But these are no ordinary strings. They are incredibly small, at 10^{-35} metres, and contain hidden dimensions.

The idea of particle strings originated in the late 1960s as physicists tried to understand how quarks were "tied" together. If quarks were thought of as the ends of little bits of elastic, rather than free particles, this could help explain why free quarks were never seen. The idea temporarily dropped out of fashion when quantum chromodynamics appeared in the mid-1970s and gave a consistent description of inter-quark forces.

However, in 1984 string enthusiasts discovered that quark strings naturally led to a theory containing supersymmetry, the ingenious scheme of pairing the known quarks and leptons with new, as yet unseen, particles. "Superstrings" were born, and looked like succeeding where field theory had failed, solving the outstanding problem in physics of the past 50 years, a marriage of quantum mechanics with general relativity – Einstein's theory of gravity.

Excitement arose when certain types of superstring theory seemed naturally free of troublesome "anomalies" – conditions where not only the predictive power of the theories but also sacrosanct conservation laws just broke down.

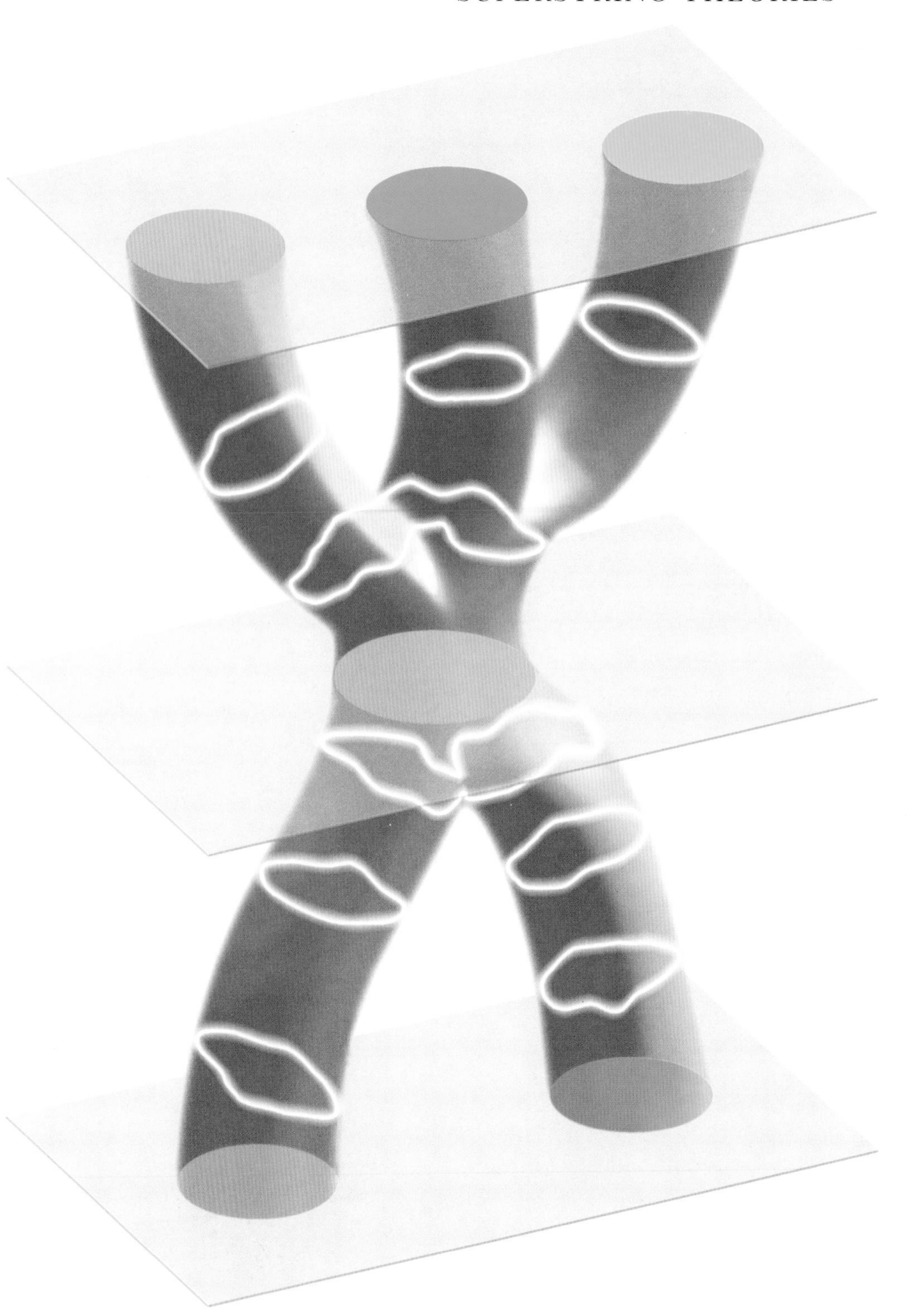

Superstrings and world sheets ***A simple particle interaction as seen on the level of strings. Two particles coming together to collide form a temporary resonance and then break up to form new particles. Instead of point particles, this picture uses closed loops containing hidden additional dimensions that we can never see. The history of these loops produces bizarre patterns of tubes. The three horizontal sheets show what is happening at successive time intervals, before, during and after the interaction.***

LASSOS IN MANY DIMENSIONS

Strings can be visualized as closed loops, like lassos, which carve out a tube in space as they move. The strings vibrate and rotate, and the quarks and leptons can be thought of as vibrations of the string, in the same way musical strings produce different notes by vibrating at different frequencies.

In practice we will never know how such strings actually behave. At about 10^{-35} metres across they are 20 powers of ten smaller than a proton, and no accelerator can or will ever be able to probe into the string regime. At this scale, quantum gravity reigns and space as we know it probably transforms into a foamy multi-dimensional structure which transcends our ordinary notion of space and time. Strings and space become interwoven in a subtle way.

The original string theory had 26 dimensions, while the superstring version has ten. For physicists accustomed to working in abstract spaces, these extra dimensions were no problem. But our everyday world has only four dimensions, three of space and one of time. Where have the six extra string dimensions gone?

They are curled up or "compactified", a process often illustrated by analogy with a hose or tube (see below). Something similar might have happened to the invisible string dimensions.

Not all string theorists are happy about compactification. "If people find the idea hard to understand and justify,"says John Ellis, former head of theory at CERN, "it's fine by me. I would prefer to formulate string theory directly into our physical world. Instead of talking about a number of dimensions some of which are rolled up, you can have a theory with a large number of internal degrees of freedom on the string, representing different quantum properties like electric charge."

PHYSICS ON ITS TOES

The superstring theory is the most promising candidate for a theory of everything (TOE), an attempt to wrap up everything in just a few equations. A TOE is a natural but ambitious goal, and requires all of the necessary physics insight to be put in place. Einstein (see page 64) knew too little of nuclear physics to be successful in his quest for a TOE. All of science is too complicated to summarize "on the front of a T-shirt", as 1989 Nobel prizewinner Leon Lederman once put it, but at least it may one day be possible to give a complete explanation of basic processes, and leave the rest to simulations on supercomputers.

"At the moment", says John Ellis, "strings are the only solution we have for reconciling quantum mechanics with general relativity, and although we lack good experimental evidence, we believe in the theory because it is consistent and beautiful. The situation in some sense resembles the early status of general relativity. Einstein's theory was motivated by theoretical consistency and not driven by experimental result."

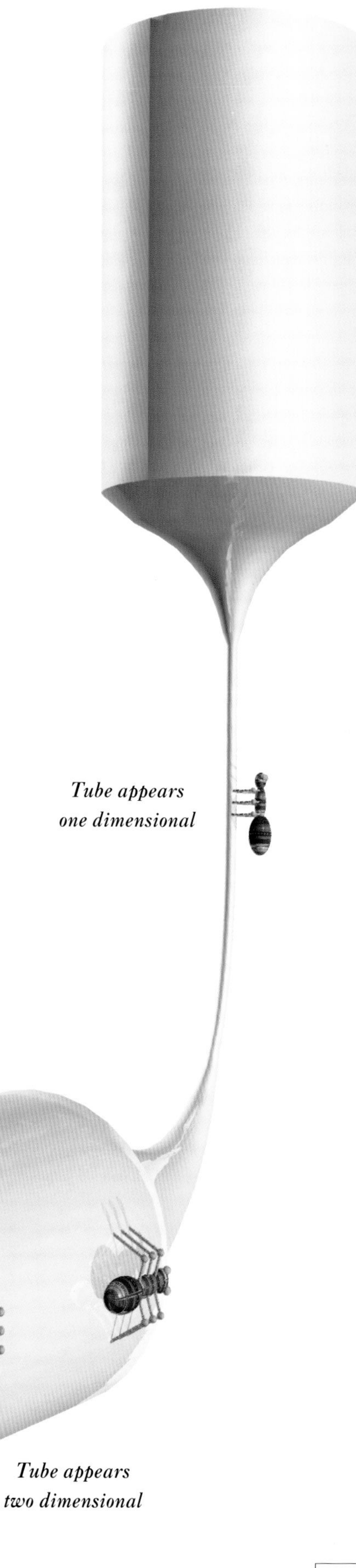

Rolled-up dimensions ***A tube is clearly a three-dimensional object. However, an ant crawling on the surface of the tube sees it as only two-dimensional. To get from one side of the tube to the other the ant has to walk around the outside of the tube; it cannot take the shortest path through the tube – the third dimension is blocked. Physicists say the third dimension is "compactified" or "rolled up". If the tube became sufficiently narrow, it would seem one-dimensional to the ant, now able to move only forwards or backwards.***

PART TWO

Looking out

From chaos to cosmos

RELATING OURSELVES TO THE UNIVERSE

Cosmology – the study of the Universe as a whole, its origin, evolution and structure – is a relatively new science. However, its history can be traced back to Ancient Greece where philosophers challenged the myths of creation and tried to make sense of the vastness around us.

In the beginning, according to the creation myths of Ancient Greece, there was "chaos", a dark shapeless void of unformed matter. The gods transformed this primordial confusion into an ordered Universe called "cosmos", a word that means "order", "harmony", or "beauty". One of the most popular myths said Light (Aether) and Day (Hemera) first formed Earth (Gaia) and Sea (Pontus). Earth then crowned creation and herself made Heaven (Uranos).

The early Greek philosopher, Thales, explained the origin of matter without invoking gods, but more or less accepted the cosmology of the Creation myths. He believed that the Earth was flat, resting on water, and said that stars were incandescent balls of vapour. His pupil Anaximander went a step further and suggested that Earth was a cylinder floating freely in space. The first to speak about a cosmos in a scientific way was Pythagoras.

MUSIC OF THE SPHERES

Pythagoras and his followers liked symmetry and claimed that the Earth was spherical. They proposed a Universe in which the Earth, Moon, and Sun, together with the five planets then known, circled a central "watchtower" called Zeus. Believing the number ten to have a special significance, they added a tenth body, a hypothetical "counter-Earth".

The Ptolemaic system ***A page from a medieval book shows the Earth at the centre of the Universe. Around the Earth are first the Moon; then two of the five planets known at the time, Mercury and Venus; the Sun comes next, then Mars, Jupiter and Saturn. Finally, the outer sphere contains the stars.***

The Milky Way *Democritus, the father of atomic theory, also had ideas about the Universe. He proposed that the Milky Way, the narrow faint band that stretches across the sky, was made up of tiny stars. His teacher, Leucippus, had claimed that stars glow because they hurtle through the air. The world "galaxy" comes from the Greek word for "milk".*

The Pythagoreans also believed that there was a deep connection between mathematics and music. The revolving planets, they said, emitted notes, the pitches of which were determined by their speed and distance from the Earth. This total harmony, "the music of the spheres", went unnoticed, they said, because it was ever-present and unchanging.

Although agreeing that the Earth was spherical, Aristotle argued that it did not move, and placed it at the centre of a spherical Universe. A pupil of Plato, he believed in pure forms, with the sphere as the basis for cosmic architecture. The other planets moved uniformly around the Earth on crystal globes, while the distant stars were embedded in an outermost heavenly sphere.

This onion-like model, first developed by Eudoxus around 370 BC, began with 27 concentric spheres. Each planet had several "shells" to account for its non-circular motion. Ptolemy in the second century AD refined this picture, which became generally accepted, acknowledged also by the Christian church. Its Universe had a radius no larger than, in today's terms, 80 million kilometres (50 million miles), only slightly larger than the orbit of Mercury round the Sun.

SUN-CENTRED UNIVERSE

Aristarchos of Samos, working in the third century BC, was the first to suggest that the Earth and planets revolve around the Sun. In his only surviving book, written when he was still young, he calculated the Sun to be 18 to 20 times the distance from Earth and size of the Moon.

Archimedes described Aristarchos' theory in his book *The Sand Reckoner*. Archimedes estimated that the Sun-centred Universe to be equivalent of about 9,000 million kilometres (5,600 million miles, or a thousandth of a light-year) across – tiny by today's standards, but considerably larger than predicted by Aristotle and Ptolemy. Like other great minds who attacked cosmic problems, Archimedes had to struggle with the deficiencies of existing mathematics for handling very large numbers and was forced to invent new schemes.

Aristarchos' radical suggestion initially did not make much headway. Challenging ancient ideas of the divine nature of the Earth and "displacing the heart of the world", it was deemed to be sacrilege. Lacking experimental support, the idea remained obscure until the sixteenth century, when it was revived by the Polish astronomer Nicolaus Copernicus, setting the stage for a new view of the Universe.

PTOLEMY – WHEELS WITHIN WHEELS

The idea of the Earth as the centre of the Universe ran into ever increasing problems as astronomical observations improved. Its greatest champion was the Egyptian Ptolemy (Claudius Ptolemaeus) in the second century AD, a skilled astronomer, philosopher and geographer. His model of the cosmos was explained in his book Syntaxis, later translated into Arabic as *Almagest*.

Ptolemy assumed that each planet moved in its own small circle – "epicycle" – as it moved around the Earth. In this way he explained how the planets appear to move against the background stars. However, to explain all details, more and more epicycles had to be added. The system became very cumbersome, but it ruled for 14 centuries.

Through the telescope

EARTH IS NOT THE HUB OF THE UNIVERSE

Early in the sixteenth century, Nicolaus Copernicus revived the idea that the Sun, not the Earth, was the centre of the Universe. When Galileo looked through the first astronomical telescope in 1609 he found that Copernicus had been right, and soon the old "geocentric" Universe idea was finally buried.

The picture of the Earth as the centre of the Universe handed down by Aristotle and Ptolemy supported the idea that people occupied a privileged position in Creation, and was acknowledged by the Christian church. During the Middle Ages this geocentric model was gradually undermined by new observations, and it became clear to the Polish astronomer and clergyman Copernicus (Niklas Koppernigk) that something was wrong.

For most of his life, Copernicus tried desperately to fit the Sun-centred Universe of Aristarchos' into the accepted picture, preserving the heavenly spheres and the circular orbits of the Ptolemaic system. Afraid of the Church's reaction, he was not keen to go into print, and the first copy of his book *On the Revolutions of the Heavenly Orbs* arrived only a few hours before he died on 24 May 1543. A friend secretly added a preface saying that the book's ideas were merely a computing device without prejudice to the truth.

On 11 November 1572, the Danish nobleman Tycho Brahe discovered a new star, shining more brightly than Venus, in the constellation of Cassiopeia. It was what we now know to have been a supernova and faded after several weeks. This was disturbing because according to classical ideas, the heavens could not change.

Tycho, who refined eyeball astronomy to its limit, also saw a comet. Carefully charting its track, he realized that comets move in celestial space and could not be products of the atmosphere as the classical ideas claimed. His observations supported a new world picture, but Tycho was not a Copernican. He firmly believed that the Sun revolved around the Earth.

GALILEO'S EYEGLASS

In 1604, Tycho's German pupil Johannes Kepler spotted a new supernova. Among the other admirers of this new bright star was Galileo Galilei, a teacher at the University of Padua. The supernova sparked his curiosity, and five years later Galileo developed the first astronomical telescope, with a magnifying power of 30 and based on recently invented military and naval telescopes.

He called his instrument an "*occhiale*", or "eyeglass", and when he pointed it towards the planets in 1609 the detailed motion of the planetary moons that he saw made him realize that Copernicus had been right. Among other things, a

The Copernican system (above) *An illustration from Andreas Callarius' book* Harmonia Macrocosmica, *published in Amsterdam, 1708. The figure on the bottom right is supposed to be Copernicus himself. Copernicus believed celestial objects moved in perfect circles and called his system the "ballet of the planets".*

Tycho Brahe's observatory (above right) *The observatory was built by Tycho at Uraniborg on Hveen, an island between Sweden and Denmark. Tycho was able to make the most accurate measurements of the stars that are possible using only the naked eye. Better measurement were made only after the invention of the telescope.*

miniature planetary system of four moons orbiting around Jupiter showed that the Earth was not the centre of all celestial operations. Galileo also looked deeper into the sky and saw that the Milky Way is a myriad of tiny stars.

Eager to broadcast his discoveries, first published in the 1610 book *The Starry Messenger*, he ran into conflict with the Roman Catholic Church. In the famous 1633 inquisition, his claims were deemed incompatible with the Scriptures, and Galileo spent the rest of his life under house arrest. Only in 1992, 360 years later, did the Church finally admit it had been wrong to condemn him.

UGLY ELLIPSES

While Galileo peered through his telescope, Kepler analysed the motions of planets, arriving at a description of planetary motion. Rather than moving in smooth circles, the planets orbited the Sun in ellipses. "I contemplate [the Universe's] beauty with incredible and ravishing delight," wrote Kepler, claiming the Pythagorean idea of celestial harmony as his inspiration. But most astronomers were not interested in his ugly ellipses.

Then, in the early 1680s, a series of large comets that were visible from Northern Europe rekindled interest in heavenly motion. In Cambridge, England, Isaac Newton had developed a theory explaining what lay behind Kepler's mysterious laws. Gravity, the attraction between bodies because of their mass, makes apples and other objects fall to the ground and controls the motions of heavenly bodies.

Newton laid out the theory in his 1687 book *Philosophiae Naturalis Principia Mathematica*, which also launched the science of mechanics, the principles of which govern motion. Central to the gravity theory is the "inverse square law". The further apart two masses are, the smaller the pull between them. In addition, it falls off as the distance multiplied by itself (squared). Thus at twice the distance the pull is four times as weak.

EDMUND HALLEY VISITS NEWTON

In August 1684, the young astronomer Edmund Halley travelled from London to Cambridge to see a recluse called Isaac Newton. Halley had been intrigued by a series of bright comets, objects that since ancient times had been considered as signs of ill omen. The comets of 1680 had alarmed Europe and were blamed for a series of disasters. In 1682 came the comet now named after Halley – shown here is an illustration from the 1493 *Nuremberg Chronicle* that depicts the AD 684 visit of the comet.

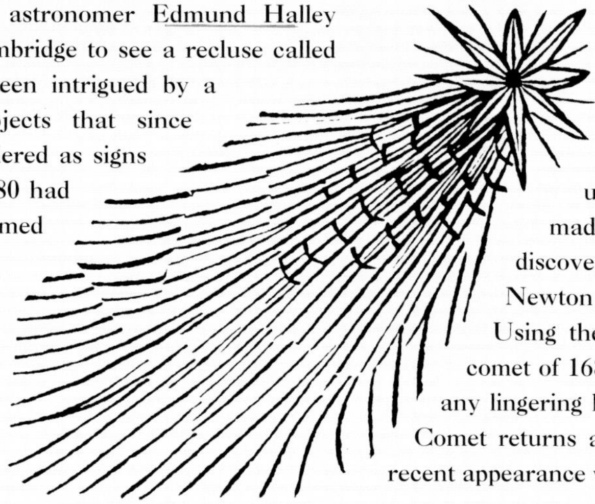

Halley desperately wanted to understand comets' orbits and hoped Newton might be able to help. Halley was astonished to find that Newton could explain Kepler's ellipses by assuming a universal gravitational attraction. He made Newton realize the importance of this discovery, and in the following three years Newton carved out his *Principia* masterpiece. Using the new theory, Halley predicted that the comet of 1682 would return in 1759. It did and killed any lingering hope for the old geocentric idea. Halley's Comet returns about every 76 years. The comet's most recent appearance was in 1986.

One galaxy among many

THE EXPANDING UNIVERSE

Early in the twentieth century, our picture of the Universe changed. The Sun is on the edge, not the centre, of the Milky Way galaxy. Our galaxy is far from being alone, and all of the galaxies are rushing apart. This revolution in astronomy was dominated by one man, Edwin Hubble.

The study of the Universe beyond the solar system was pioneered by the great eighteenth-century astronomer Sir William Herschel. In 1784 he concluded that the Sun was one of many central stars in a vast disc-shaped galaxy. The faint band of light, known since ancient times as the "Milky Way", was the galaxy seen sideways.

Herschel also saw many faint cloud-like patches, or "nebulae", which seemed to be far outside the Milky Way. Some looked to be spiral and puzzled astronomers. In 1755 the German philosopher Immanuel Kant suggested they could be separate galaxies, which he called "island universes". Others believed nebulae were clouds of gas inside the Milky Way that were condensing into new stars.

In 1901 the Dutch astronomer Jacobus Kapteyn completed the first map of the Milky Way, estimating its diameter to be 23,000 light-years. Many astronomers still believed that our Milky Way dominated the heavens. Beyond this might be an infinite starless void, but this was of no interest. If by any chance it contained stars, many astronomers believed there was no hope of ever seeing them.

Better measurements of the size of the galaxy first needed an improved method of judging the distances of stars. Using variable stars called Cepheids as a kind of astronomical yardstick (see page 138), the American astronomer Harlow Shapley charted the Milky Way. He discovered it was circular, with the Sun near its outer edge, rather than in the centre. Four hundred years after the Copernican revolution, the focus shifted again; this time the Sun was no longer thought to be the centre of the Universe.

Shapley still believed that the Milky Way was the only galaxy in the Universe. However, the island universe picture was gaining ground, and in 1924 the controversy was solved when Edwin Hubble found Cepheids in Andromeda, showing that it was a separate star system beyond the Milky Way, and not just a gaseous nebula. Andromeda is 2.5 million light-years away and the furthest object

Herschel's 40-foot telescope (above) *Herschel built his 12-metre long telescope in Windsor, England, and it drew many visitors. It had an elaborate system of supports and ladders. The observer stood on a platform at the mouth of the telescope.*

Close neighbour (right) *Andromeda (M31), identified by Edwin Hubble in 1923 as the nearest galaxy beyond our own.*

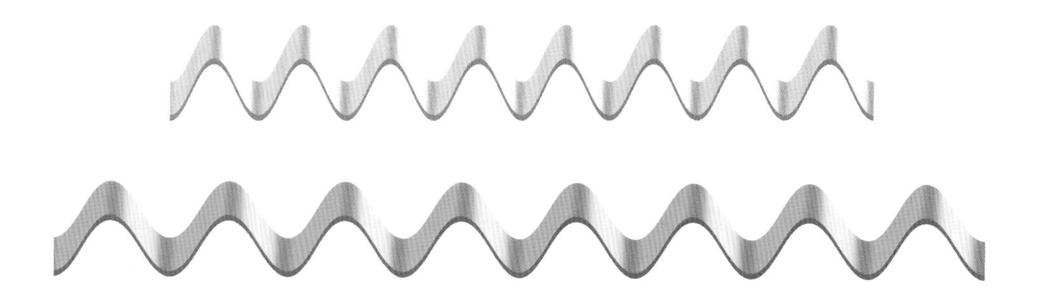

Red shift *The expansion of the Universe causes light from distant galaxies to be stretched towards the red end of the spectrum. Light from Andromeda, however, is squeezed because of motions in our local galaxy group and is bluish in colour. Shifts due to local movement are analogous to the Doppler effect in sound.*

visible with the naked eye. It is a spiral galaxy very similar to our Milky Way and contains about 200,000 stars.

Soon observations showed that certain other nebulae contained stars and were galaxies in their own right. Suddenly astronomers realized that the Universe was much bigger than they had believed. But there was another surprise in store.

GALAXIES ON THE RUN

In 1914, Vesto M. Slipher saw that light from distant nebulae looked redder. Hubble realized this could be because the Universe is expanding.

The wavelength of light, its "stride" measured by the distance between successive wave crests, increases along with the rest of space as the Universe expands. The longer the wavelength, the redder the light. The degree of increase in redness shows how much the stride has lengthened – how much the Universe has expanded – since the light was emitted. The rate of expansion of the Universe looks like a velocity: the further away an object is, the more the intervening space has expanded and the faster the object appears to recede. Plotting a graph of velocity against distance for 24 galaxies, Hubble found a straight line – Hubble's law – which provides a way of estimating cosmological age (see page 138).

In subsequent years, Hubble and his assistant Milton Humason extended their survey using, as Hubble had before, the 100-inch Mount Wilson telescope in Pasadena, California. By 1935 they had charted galaxies as much as 100 million light-years away, a much greater distance than that believed to be the dimensions of the Universe in 1900.

EDWIN HUBBLE – THE HEAVYWEIGHT

Edwin Powell Hubble was born in Marchfield, Missouri, in 1889, the son of a local lawyer. He graduated from Chicago University in 1910, where he had studied both law and astronomy. He also distinguishing himself as a heavyweight boxer and was invited to turn professional, but turned down the offer. Later he fought an exhibition match with the French champion Georges Carpentier.

In 1913 Hubble worked as lawyer for a few months, but returned to astronomy. He started his historic career in 1919 after serving as major in France during World War I.

A day without a yesterday

POINTING BACK IN TIME

Cosmology, the science of understanding the Universe, was revolutionized in 1927 by a Belgian priest and astronomer who put forward the first modern hypothesis of creation. Supported by Hubble's discovery of the expanding Universe, it eventually led to the idea of the "Big Bang".

It is difficult to think of the genius of Albert Einstein making a monumental mistake. It happened in 1917, however, when he published his famous paper "Cosmological considerations on the general theory of relativity". His calculations showed that the Universe was curved by gravity into a closed four-dimensional sphere with a diameter of about 100 million light-years.

Einstein's mathematical model of the Universe had a disturbing property – objects in it, such as galaxies, were moving apart or falling together. Not knowing what Hubble would later discover, this was contrary to Einstein's convictions. Like most scientists of the time, he believed the real Universe to be static and unchanging. To stop the galaxies from moving he added an extra piece to his equation, calling it the "cosmological constant", in effect a force that worked against gravity.

EINSTEIN'S BIG BLUNDER

The extra piece that Einstein added was a mathematical cheat, a fictitious anti-gravity mechanism that pushed matter apart rather than pulling it together. Although it made his results look more conventional at the time, he never liked it. It destroyed the formal beauty of the equations – always a strong point of his work. Later he admitted the cosmological constant to be "the biggest blunder of my life". If he had accepted what his beautiful equations told him, Einstein would have predicted the expanding Universe more than a decade before Hubble discovered it.

In 1917, the Dutch astronomer Willem de Sitter put forward a solution to Einstein's equations. Retaining the false cosmological

Creation *An artist's impression of the Big Bang. A remarkable musical description of the Big Bang is found in Joseph Haydn's oratorio* The Creation, *says physicist Victor Weisskopf. The choir sings softly, "Let there be light", then bursts into a C-major chord. "There can not be a more beautiful and impressive artistic rendition of the beginning of everything."*

MONSIEUR L'ABBÉ – GEORGE LEMAÎTRE

Georges Lemaître was born in Charleroi, Belgium, in 1894. After active service in the First World War, he quickly fulfilled two great ambitions. After obtaining a rapid Ph.D. in mathematics in 1920, he joined a seminary to train for the priesthood and was ordained in 1923. He then turned to research in astrophysics and cosmology at Cambridge under Sir Arthur Eddington.

Lemaître's revolutionary ideas were published as a collection of essays in 1933. He became a scientific celebrity, but shunned the limelight and spent the rest of his life studying other research problems. In particular he thought cosmic rays were evidence for his theory, with their origin in the primordial explosion.

In 1965 he suffered a heart attack from which he never recovered, and while lying in hospital received the 1 July issue of the *Astrophysical Journal,* which contained news of the discovery of a cosmic background radiation (see page 100), confirming the Big Bang idea. He died the following year, aged 72.

constant, he arrived at a result that seemed absurd: de Sitter's Universe was empty and could only remain static as long as it contained no matter. Introducing objects, such as two galaxies, into the model, others found a strange effect. The objects started to fly apart.

From 1922 to 1924 in Leningrad, the Soviet mathematician and meteorologist Alexander Friedmann threw away the cosmological constant and produced several different solutions to Einstein's relativity equations, all showing that the Universe expanded. He pointed out that Einstein had made a mathematical error, dividing by a quantity that could be zero and therefore introduce infinities.

The relatively unknown Friedmann plucked up courage and wrote to Einstein pointing out the mistake. The world-famous scientist had the humility to admit that Friedmann was correct. However, the Russian did not live long enough to see the final triumph of his ideas. He died of typhoid in 1925 at the age of 37.

PRIMEVAL ATOM

In 1927, the obscure Belgian astronomer and priest Georges Lemaître presented a theory of an expanding Universe that had a beginning in time – "a day without a yesterday", as he called it. Although apparently ignorant of Friedmann's work, he suggested that the Universe was born in a primordial explosion. This idea was published in a little-read Belgian journal and went unnoticed until Hubble's discovery of the expanding Universe.

Lemaître proposed the initial Universe as a highly compressed state of matter, which he called the "primeval atom" (*l'atome primitif*), a sort of giant superheavy neutron that disintegrated by some kind of radioactivity. This was the opposite of today's view, where the Universe first builds the simplest atomic nuclei before going on to make larger ones. Although Lemaître's ideas differed from modern cosmology in several other respects, he is rightfully looked on as the father of the Big Bang theory.

In with a bang

ACCEPTING THE BIG BANG

At first the idea of a Big Bang did not catch on. The preferred picture was a "steady state" Universe, in which matter is continually being created. The verdict came in 1965 when a faint but persistent radio hiss told scientists they were listening to the distant echo of the Big Bang.

The modern version of Lemaître's idea of a giant primordial explosion was worked out in the late 1940s by the eccentric physicist George Gamov (once a student of Alexander Friedmann), assisted by Ralph Alpher and Robert Herman.

Gamov and his colleagues realized that if the early Universe had been a dense fireball, protons and neutrons would have stuck together to form the nuclei of atoms. Initially they thought all possible nuclei would have been formed this way, but calculations showed that the nuclear production line of the early Universe would have stopped at helium, which is an extremely stable nucleus with two protons and two neutrons held very tightly together. So after making its helium, the Universe had paused.

HELIUM PROBLEM

The primordial helium locked up in stars is still one of the major ingredients of the Universe, making up about a quarter of its total weight. On Earth, only traces of helium are found, and the element was discovered in the nineteenth century by analysing sunlight, hence its name, from the Greek "*helios*" meaning "Sun".

In 1948, the year that Gamov published his work on what eventually became known as the Big Bang picture, Fred Hoyle, Hermann Bondi and Thomas Gold in England, who had worked together on radar development during the World War II, put forward a totally different idea. The Universe had to expand, but they rejected the idea of an initial explosion. The Universe, they said, has always been in a "steady state", without a beginning, and without there having to be an end. They

suggested that new matter – in the form of hydrogen gas – is continuously being created in the space between galaxies as they move apart.

The two theories clashed for over a decade, with continual confrontations between rival protagonists. During this controversy Fred Hoyle, on BBC radio, facetiously coined the name "Big Bang" for the rival picture. Committed to their idea, Hoyle and his co-workers realized that a range of nuclei, and not just helium, could be cooked deep inside stars.

By the early 1960s, detailed analysis of starlight showed that while about 75 per cent of the Universe was made up of hydrogen, almost all the rest was helium. The presence of much helium and so few other nuclei was difficult to explain in the steady-state picture. But helium was good news for the Big Bang.

RESIDUAL RADIATION

Gamov had also realized that the heat of the Big Bang should still linger. However, this radiation would have been so stretched and enfeebled by the expansion of the Universe that it would appear to very be cold, just a few degrees above absolute zero – the point at which molecules of matter cease to move. This cold cosmic background radiation, he said, should be detectable.

For 15 years the prediction lay forgotten. In 1964, a group of Princeton physicists, led by Robert Dicke, began thinking along similar lines and built a radio telescope to look for background radiation. Although their theory was slightly different, they reinvented Gamov's result; a dim, cold glimmer would be all that remained of the initial fireball.

While Dicke and his group were preparing for their search, Arno Penzias and Robert Wilson were busy nearby at the Bell Laboratories, New Jersey, with a horn antenna, developed for receiving signals from *Telstar* (the first transatlantic television and communications satellite) but converted for radio astronomy. Penzias and Wilson did not know about Gamov's prediction and were not particularly interested in the birth of the Universe.

They became puzzled by a faint radio hiss, or "excess noise", as they called it, which would not go away. This hiss was in the microwave part of the spectrum, with wavelengths shorter than radio waves but longer than infra-red. It corresponded to a temperature only a few degrees above absolute zero, and was always the same, no matter where the antenna pointed. After checking their equipment, and even cleaning off droppings from two nesting pigeons, they realized this microwave radiation had to come from outer space.

Having heard about Dicke's project, Penzias and Wilson contacted the Princeton team, who immediately saw the significance of the discovery. The two groups discussed their results and their papers were published simultaneously in the *Astrophysical Journal*. Penzias and Wilson received the Nobel Prize for Physics in 1978.

The New Y

No. 39,199. © 1965 by The New York Times Company. Times Square, New York, N. Y. 10036 NEW YORK, F

Signals Imply a 'Big Bang' Univers

Horn antenna, used in space exploration, at the Bell Laboratories in Holmdel, N. J.

By WALTER SULLIVAN

Scientists at the Bell Telephone Laboratories have observed what a group at Princeton University believes may be remnants of an explosion that gave birth to the universe.

These remnants are thought to have originated in the burst of light from that cataclysmic event.

Such a primordial explosion is embodied in the "big bang" theory of the universe. It seeks to explain the observation that virtually all distant galaxies are flying away from the earth. Their motion implies that they all originated at a single point 10 or 15 billion years ago.

The Bell observations, made by Drs. Arno A. Penzias and Robert W. Wilson from a hilltop in Holmdel, N. J., were of radio waves that appear to be flying in all directions through the universe. Since radio waves and light waves are identical, except for their wavelength, these are thought to be remnants of light waves from the primordial flash.

The waves were stretched into radio waves by the vast expansion of the universe that has occurred since the explosion and release of the waves from the expanding gas cloud born of the fireball.

In what may prove to be one of the most remarkable coincidences in scientific history, the existence of such waves was predicted at

Continued on Page 18, Column

KENNEDY CRITICAL | Jersey Central Asks | PLAN IS SPEEDED

A discovery goes public (above) *On 21 May 1965, the* New York Times' *front page led with the announcement that "Scientists at the Bell Laboratories have observed what a group at Princeton University believes may be remnants of the explosion that gave birth to the Universe."*

Hearing the Big Bang (left) *Arno Penzias (on the left) and Robert Wilson revisiting the horn-shaped antenna at the Bell Laboratories where in 1964 they by chance detected the faint hiss of the cosmic background radiation in all directions in the sky. Their discovery was strong evidence in support of the Big Bang theory and is one of the milestones of twentieth-century science.*

ALPHER, BETHE, GAMOV

Although he grappled with the hardest problems of physics, Gamov could not spell or do arithmetic. Born in Russia in 1904, he moved to the United States in 1933. As well as doing basic research, he wrote a series of science books where "Mr Tompkins" was the bemused central figure. An inventive mind, he called the original stuff of the Universe "ylem", from a Greek word meaning "primordial matter".

Working with his student Ralph Alpher in Washington during the mid-1940s, he learned that the German physicist Hans Bethe had arrived in New York. Gamov invited him to join them for a famous paper on nuclear transformations, known as "Alpher, Bethe, Gamov", alluding to the first three letters of the Greek alphabet.

A very special bang

GETTING THE INITIAL CONDITIONS RIGHT

After initial successes, the Big Bang idea ran into serious difficulties. The conditions of creation had to be very special; it was not just any old bang. A major rethink rewound the picture much closer to time zero and led to new links between cosmology and particle physics.

All through its history, the Universe has pushed outwards against the inward pull of the gravity of its own mass. In somewhat the same way, all our lives we have to combat the gravity of the Earth. The only way to escape Earth's gravity is to jump. If the jump is powerful enough, an object has sufficient energy to sail out into space. If the jump is too weak, the object falls back. At a finely tuned "critical" jump, it does not completely escape, but coasts in a captive orbit.

For the Universe, the force of the Big Bang explosion provided a jump. If this was bigger than the critical level, the Universe would be "open", expanding for ever, with the galaxies getting further and further apart. On the other hand, if the Big Bang was sub-critical, without enough power to overcome gravity, the Universe would be "closed", ultimately collapsing in a "Big Crunch" – a Big Bang in reverse.

This has not happened despite our Universe having existed for some 15,000 million years (see page 138), suggesting that we are very near the critical level. To maintain such a precarious position – neither exploding rapidly outwards nor collapsing inwards – for so long, like balancing a huge pyramid on its point, means the Big Bang conditions must have equalled the critical level to an accuracy of 50 decimal places. Anything less, and the Universe would long since have expanded out of sight; more and it would have collapsed back on itself.

FAR HORIZONS

Another problem for the Big Bang was the eerie uniformity of the background radiation. To a good approximation, this has the same temperature in all parts of the sky. At any given time in the Universe's history, there is a limit – the "horizon distance" – to how far light can have travelled since the Big Bang. For us, the horizon distance marks the edge of the observable Universe, and beyond it is more Universe that we have not yet seen – its light is still on its way to us. All the time, galaxies from the outer regions of the expanding Universe are creeping over this horizon and into view.

Going back in time, this horizon shrinks because light has had less time to travel. When the background radiation escaped, what are now opposite sides of the sky were then separated by over 90 horizon distances. For this radiation to "know" that it had to have almost the same temperature everywhere, either it must have communicated at the impossible rate of 100 times the speed of light, or something else happened.

In 1978, Alan Guth, a young American particle physicist working at Cornell University, was listening to a lecture by Robert Dicke about the Big Bang. Guth knew little about cosmology at that time; his interest was a single "grand unified theory" (GUT, see page 84). After the talk, he began to think about the implications of these ideas for cosmology.

Light cones (right) ***Looking back in time along a ray of light from any point gives the largest region of space that can be connected to an observer. This region of space is illustrated by "light cones", which show that distant parts of this space were too far apart to have been connected in the past and could not have influenced each other because the Big Bang was not an infinite time ago.***

Inflation (below) ***A quantum bubble creates space in a supercooled Universe and expands millions of millions of millions of times faster than the speed of light. Towards the end of inflation, surplus energy is dumped into space, reheating the Universe and giving birth to new matter.***

ALAN GUTH'S SPECTACULAR REALIZATION

Alan Guth, now Professor at the Massachusetts Institute of Technology, was born in 1947. Inflation "just seemed to fall into my lap", he said; "all the important information was there".

On 6 December 1979, he came home from a day's work at the Stanford Linear Accelerator Center and sat down to do a few more calculations. He kept working into the small hours. The following morning he rushed back into work, brandishing a red notebook full of jottings. At the top of the page was the headline "Spectacular realization". It was the sort of breakthrough every research scientist dreams of.

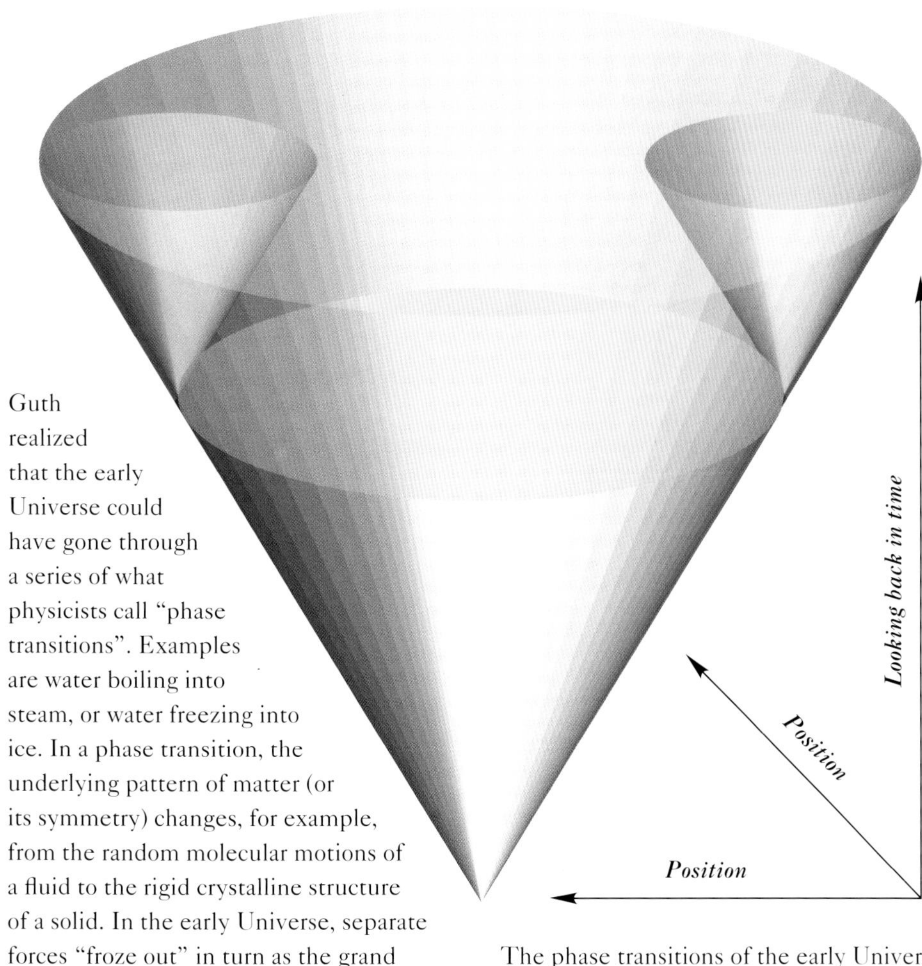

Guth realized that the early Universe could have gone through a series of what physicists call "phase transitions". Examples are water boiling into steam, or water freezing into ice. In a phase transition, the underlying pattern of matter (or its symmetry) changes, for example, from the random molecular motions of a fluid to the rigid crystalline structure of a solid. In the early Universe, separate forces "froze out" in turn as the grand unification force (see GUT, page 84) gradually cooled.

Phase transitions normally happen quickly. However, water, if it is cooled slowly and carefully enough, can stay liquid despite being up to 20 degrees below its usual freezing point. This is called "supercooling". Something similar might have happened to the early Universe, where new force conditions should have "frozen in" at definite levels, but the old conditions somehow managed to survive.

The phase transitions of the early Universe also happened very quickly. At just 10^{-35} second, the Universe-dot had "cooled" below 10^{17} degrees. The strong nuclear force should have "frozen out", but the Universe, or some part of it, remained in its earlier state, with its underlying symmetry supercooled.

However a tiny quantum bubble (see page 36) managed to seep out into the surrounding vacuum. As this special bubble expanded, it created new space with its own energy density, thus trying to shake off this relentless accumulation of energy, it expanded faster than anything else the Universe has ever known, faster even than light. Termed "inflation" by Guth, this made the bubble increase by a factor of 10^{50}, doubling its diameter in each 10^{-34} second – one millionth of the time it takes for light to cross a quark. The tiny insignificant bubble turned into the biggest thing around by far.

Finally, the supercooled strong-force region "remembered" that it was unstable, and the region froze. The surplus energy made in the meantime was dumped, reheating the Universe to 10^{27} degrees and making many more particles. After this, the Universe reverted to its much slower Big Bang expansion, and fresh stages of freezing out set the stage for the development of the Universe as we know it today.

The sudden puff of inflation magically wiped out the puzzles of the initial Big Bang cosmology. The visible Universe evolved from a tiny region of space where energy was uniformly spread out. All parts of this minute "sky" were momentarily in light-ray range before being irrevocably blown apart. Whatever it was before, the density of the early Universe was quickly pushed to the "fine-tuned" critical level, and has stayed there ever since.

But this new inflation picture still had a flaw, sometimes called the "graceful exit problem". The Soviet cosmologist Andrei Linde put forward another idea, with chance quantum fluctuations giving concentrations of energy acting against gravity and pushing the Universe outwards. Such "chaotic inflation" could happen over and over again, giving what Linde calls an "eternally existing self-reproducing chaotic inflationary Universe"!

The biggest explosion ever

THE BIG BANG

10^{-43} sec *Gravity separates as a force.*

10^{-32} sec *Inflation ceases; the Big Bang expansion continues.*

10^{-11} sec *Electroweak force splits into electromagnetism and the weak force.*

Our Universe was born about 15,000 million years ago in a titanic explosion, the Big Bang. A pinpoint of superdense and intensely hot matter erupted in a fierce burst of energy creating even space itself, which is still expanding.

The first second in the life of the Universe sets the scene. Insignificantly short by ordinary standards, this initial epoch was packed with major cosmic events.

10^{-43} second: the action begins After a brief prologue, space and time begin to have a meaning. At 10^{32} degrees the Universe, a tiny point 10^{-32} centimetres (inches) in diameter containing an exotic mix of particles and antiparticles being created and annihilated, sees its first piece of history: gravity separates out as a force on its own. This separation is one of the "phase transitions" in which the forces in the Universe "freeze out" from a unified force as temperature drops.

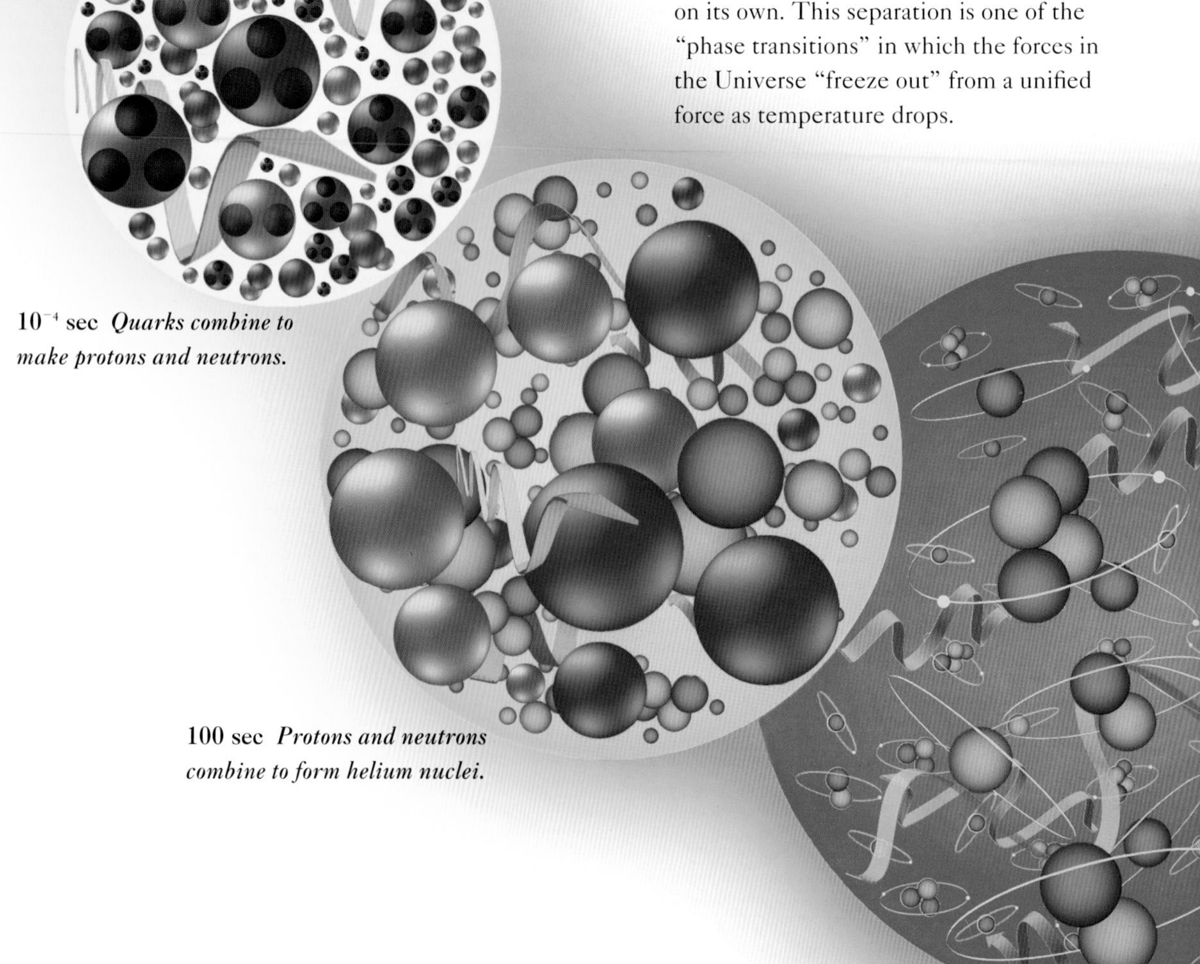

10^{-4} sec *Quarks combine to make protons and neutrons.*

100 sec *Protons and neutrons combine to form helium nuclei.*

300,000 years *The Universe becomes transparent and fills with light.*

KEY

- *Exotic particles*
- *Photons*
- *Quarks*
- *Electrons*
- *W-particles*
- *Z-particles*
- *Mesons*
- *Neutrons*
- *Protons*
- *Nuclei*
- *Atoms*

Positively charged particles are shown as blue, negatively charged ones as yellow and neutral ones as grey. Once the quarks combined to form protons, neutrons and mesons, they became invisible and thus are not shown in later stages.

10^{-35} second: inflation begins As the strong force tries to freeze out, quantum bubbles leak into the surrounding vacuum. One bubble inflates at enormous speed; inside it our visible Universe reaches about the size of a tennis ball. All the forces (except gravity) are still united, but the symmetric vacuum suddenly "realizes" it is unstable and pours off its energy, producing more particles, and the strong force freezes out. (Inflation is described in more detail on page 102.)

10^{-32} second: inflation stops The much slower but still mighty Big Bang expansion takes over again. There are two kinds of particle: quarks, which feel the strong force, and leptons, which feel the still-unified electroweak force.

10^{-11} second: the electroweak split The temperature drops below 10^{15} degrees, another freezing point. The electroweak force splits into electromagnetism and the weak force in a symmetry breaking process (see page 66). The weak force carriers, W and Z, are heavy, while the photon, the carrier of electromagnetism, has no mass.

10^{-6} second: quark massacre Quarks and antiquarks have been flying around, being created, annihilating each other and interacting with other particles. As the Universe cools to 10^{13} degrees, there is no longer enough energy for quarks to be produced spontaneously. The pairs in existence continue to annihilate, and it looks as though quarks will vanish for ever.

10^{-4} second: baryons form The Universe has grown to about the size of our Solar System. As the temperature falls, quark annihilation stops and remaining quarks combine to make protons and neutrons.

1 second: the great neutrino escape The neutrinos, which feel only the weak force, have so far been very active. But at the end of the first second the weak force has become so weak that it has virtually no hold on the neutrinos any more, and they fly off on their own. They are still "out there" in vast numbers.

100 seconds: the first elements Protons and neutrons suddenly combine to form helium nuclei. Nothing much happens for the next 100,000 years or so. Hydrogen, helium and traces of a few other light nuclei, mixed with electrons and radiation, gradually cool to about the temperature of a steel furnace.

300,000 years: the Universe lights up Electrons begin to stick to nuclei. The first atomic matter is born. Radiation, no longer strong enough to break atoms apart and is not automatically absorbed. The Universe becomes transparent and is filled with light.

1,000 million years As galaxies form, the Universe takes on a more familiar shape.

15,000 million years The Universe as we know it, on both cosmic and atomic scales.

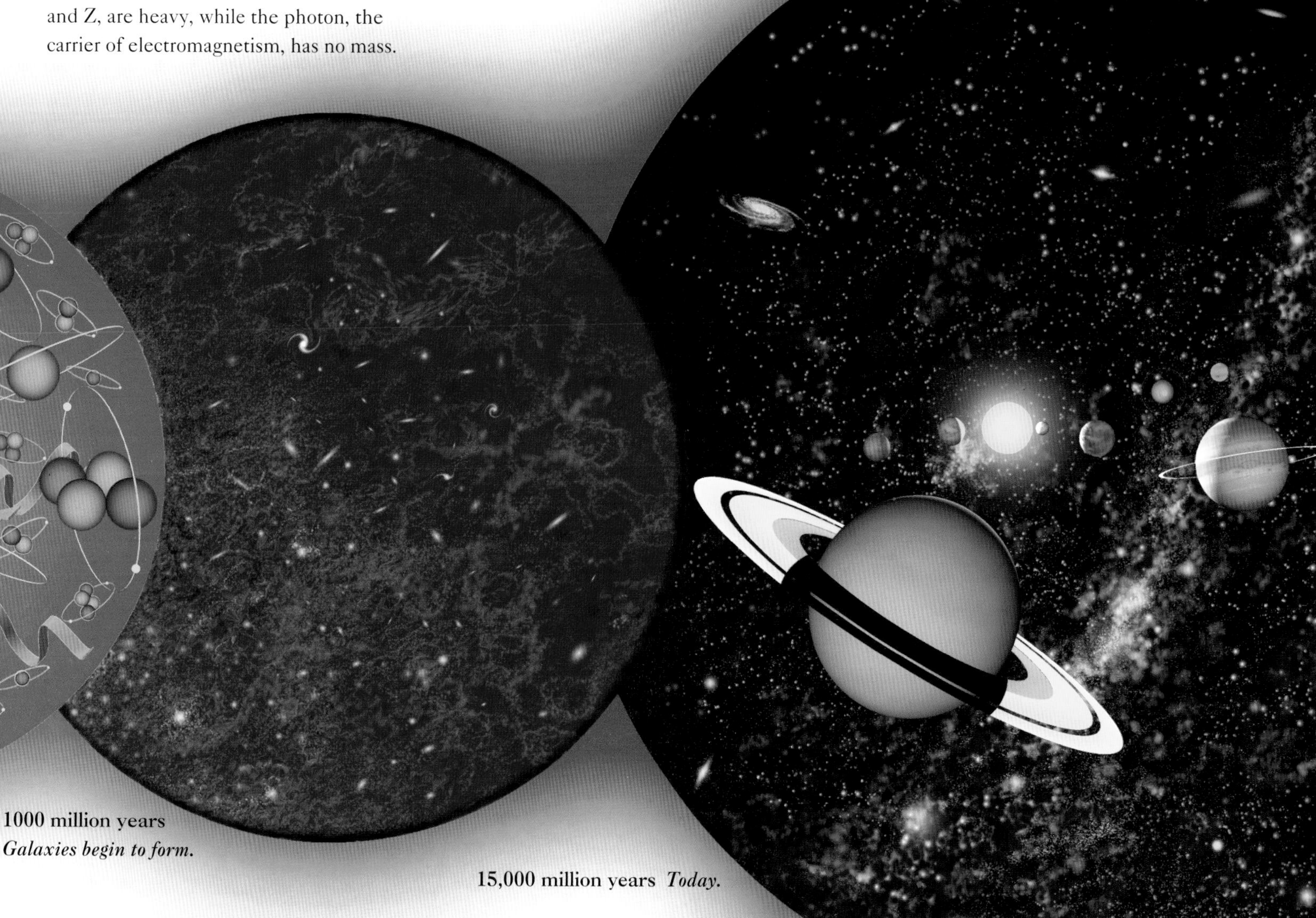

1000 million years *Galaxies begin to form.*

15,000 million years *Today.*

Star-spangled backdrop

NUCLEAR PHYSICS HELPS EXPLAIN THE STARS

Stars – tiny luminous spangles on the cold dark fabric of the Universe – are born when wisps of gas and dust in interstellar space are pulled together by gravity. It was not until the 1930s, when physicists discovered how the atomic nucleus works, that astrophysicists understood what makes stars shine.

Twinkling in the sky, stars seem eternal. But on the vast time-scale of the Universe they have a limited life. When their nuclear fuel is used up, some fade away gently, others suffer a violent death. Stars are scattered at enormous distances. The nearest, apart from the Sun, is Proxima Centauri, 4 light-years away. In the late nineteenth century, William Thomson (Lord Kelvin) in Scotland and Hermann von Helmholtz in Germany supposed that a star was a cloud of gas that had contracted under its own gravity, heating up by friction and eventually becoming luminous. But this way the Sun could only live for 20 million years or so, a figure difficult to reconcile with the geological history of the Earth, where time spans are measured in thousands of millions of years. If the Sun is at least as old as the Earth, where does its durable power come from?

Thirty years later, the British astronomer Arthur Eddington guessed that the Sun gets its energy by somehow burning hydrogen into helium, but could not explain why. This was left for two German physicists, Hans Bethe (working in the United States) and Carl von Weizsäcker in the 1930s. Understanding the stars needed the new insights of nuclear physics.

The inner furnace of a star like the Sun is kindled by a very difficult and rare nuclear spark. An average stellar hydrogen nucleus (a proton) bounces around for thousands of millions of years before finally dodging the electrical repulsion of another proton and fusing with it under the gentle action of the weak force (see page 48). However, stars have a vast supply of protons so, by the laws of chance, many of

ARTHUR EDDINGTON – EINSTEIN CONFIRMED

The British physicist and astronomer Sir Arthur Eddington (1882-1944) was one of the first to see the importance of Einstein's theories of relativity. He planned an expedition to observe the 1919 solar eclipse, only visible in the tropics. The observations showed light from distant stars bent by the Sun's gravity and put Einstein's name into the headlines.

White dwarf in the Helix Nebula (above) *Astronomers have catalogued some 1,600 planetary nebulae, spherical shells of gas formed as dying Sun-like stars fling their outer layers into space. The largest and most beautiful is the Helix Nebula, 500 light-years away in the constellation of Aquarius. Helix is expanding at about 30 km (20 miles) per second. Ultraviolet radiation from an intensely hot white dwarf, the remnant of the original star at the centre of the nebula, lights up the peripheral gas cloud.*

Red supergiant in Orion (left) *There are several red supergiants, stars that have swollen and are on their way out. One is Betelgeuse in the left shoulder of Orion the Hunter (seen here in an enhanced photograph), about a thousand times bigger than the Sun and 60,000 times brighter.*

these interactions occur. After several more fusions, the result is helium, each nucleus of which weighs less than the sum of its component particles. This mass difference is released deep in the star as $E=mc^2$ energy (see page 40), radiating outwards towards the surface.

The outward pressure of this "starshine" counters the inward pull of gravity, and as long as it has hydrogen at its core to burn, the star is stable. When the fuel supply is exhausted, the inner thermonuclear furnace cuts out and the pressure drops. Gravity re-exerts its grip and the star begins to collapse.

The Sun, born some 4,600 million years ago, burns about 1,000 million ton(ne)s of hydrogen into helium each second, but should shine for about as long again before it runs out of fuel.

RED GIANTS, WHITE DWARVES

The fate of a star is determined by its mass. When an a star of average size – such as the Sun – starts to collapse, its compressed core gets hotter, igniting helium "ash" from the original thermonuclear furnace and making carbon. The extra heat forces back the surrounding envelope of gas, ballooning the star out to hundreds of times its former size. With its radiant energy spread over a much larger surface, the star fades to become a "red giant".

When its nuclear fuel is spent, a red giant's outer envelope, frayed by the wind from the stellar interior, blasts millions of ton(ne)s of material out into space, forming an immense hollow shell of gas called a planetary nebula. At the centre of the nebula, the remnant of the original star shrinks to a "white dwarf", about the size of the Earth but a million times heavier.

The white dwarf is rescued from further collapse by quantum physics. As the core is compressed, its atoms are squashed – the space between nuclei and the surrounding electrons is squeezed out. But a quantum-mechanical limit (Pauli's exclusion principle) on how tightly electrons can be packed counters the crush of gravity.

A white dwarf is initially very hot, lighting up the inside of the surrounding nebula. But with no energy supply the star inexorably fades, dying with a whimper.

At the beginning of the century, most astronomers thought that white dwarves were the ultimate stellar tombstones. However, in 1931 the young astrophysicist Subrahmanyan Chandrasekhar, on his way to England from India, realized that when burnt-out stars reach about 1.5 times the weight of the Sun, the crush of gravity overcomes even the resistance of the exclusion principle. The resulting nuclear pulp is the raw material for supernova explosions and other cosmic fireworks.

Many stars leave small remnant husks which sit quietly in space, accumulating dust. This, or a chance endowment of extra matter, can tip the star's mass over the "Chandrasekhar limit", producing a new supernova explosion.

From ashes to dust

OLD STARS MAKE NEW MATTER

On 23 February 1987, a star appeared to explode in our neighbour galaxy the Large Magellanic Cloud, 170,000 light-years away. The brightest supernova for nearly 400 years, it gave a rare opportunity to prove that the cataclysmic death of big stars is the birthplace of heavy nuclei.

Just two nuclei, hydrogen and helium, were produced from the Big Bang and provide the basic cosmic material. Although some of them are plentiful on Earth, the other 90 naturally occurring nuclei together make up no more than 1 per cent of the Universe. Thirty years ago astrophysicists thought that all but a few nuclei were produced in a chain of nuclear reactions inside stars. Later they realized that stellar fusion could not produce nuclei larger than that of iron, number 26 in the periodic table (see page 20). Further nuclear reactions could only be sparked off when big stars exploded, providing a sudden boost of extra energy.

VIOLENT DEATH

A star over ten times bigger than the Sun is fated from birth to a violent end as a supernova. Its immense gravitational crush pushes its core temperature up to 600 million degrees, turning it into a huge pressure cooker that squanders its core fuel. While smaller stars fuse hydrogen into helium and then helium to carbon, ending as a burnt-out "white dwarf", bigger stars continue to contract and heat up, tripping new thermonuclear switches to roast carbon into successively heavier nuclei, such as neon, oxygen and silicon.

After silicon has been roasted into iron, the fusion route is suddenly blocked. Iron is the most stable nucleus – fusing nuclei lighter than iron releases energy, but energy has to be fed in to form heavier ones. With its energy supply suddenly cut off, the core of the star collapses in less than a second, its mantle imploding on the core, crushing it tightly. This final implosion provides an extra burst of energy that transforms even the cinders of the star, superheating pockets of iron into the whole range of heavy nuclei.

The core material has already been squeezed so much that it springs back into shape like a rubber ball, sending out a powerful shock wave that blasts off the star's outer layers, flinging the stew of nuclei far out into space. The kernel settles as a compact neutron star (see page 110). Very massive stars collapse even further and become black holes (see page 116).

The nuclear-rich stellar dust is the raw material for new cosmic projects. Earth formed 4,600 million years ago after one or two supernova explosions had brewed all the necessary ingredients, including those for the human body. With oxygen making up 65 per cent of the body and carbon 18 per cent, together with minerals such as phosphorus, we are mostly made of old stardust.

FRITZ ZWICKY AND OBJECT HADES

In 1934 Fritz Zwicky suggested that the massive explosions caused by the collapse of a neutron star could produce supernovae. Together with Walter Baade at the California Institute of Technology he searched for these mighty eruptions, discovering about a hundred. Today hundreds have been recorded and catalogued.

Zwicky, born in Bulgaria of Norwegian parents, grew up in Switzerland. In the United States, his Caltech colleagues called him the "Mad Swiss" because of his propensity for toying with far-fetched ideas. He dared to suggest that a neutron star was not the end of the road and collapse could go further, producing what he called an "object Hades" – what came to be known as a black hole (see page 116).

A supernova appears *This pair of photographs was taken before and after the explosion of the supernova SN1987A. The upper was taken in 1969 and the lower on 26 February 1987, two days after the explosion. Supernova 1987A was the brightest seen since 1604 and resulted from the explosion of a giant star in the Large Magellanic Cloud, a galaxy near the Milky Way. This windfall cataclysm gave astrophysicists a golden opportunity to follow the fate of the Supernova and test their understanding of stellar dynamics. It was the first time a supernova had come under such detailed scrutiny.*

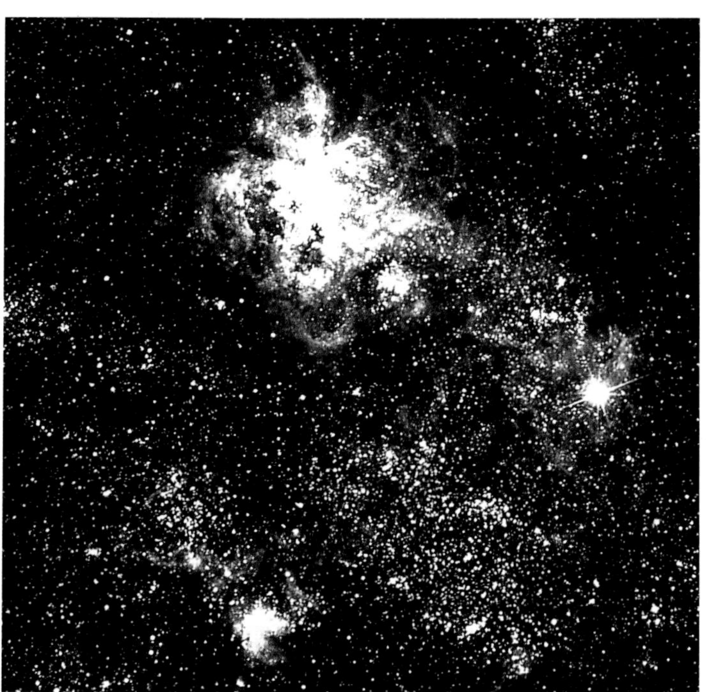

BLUE SUPERGIANT

Supernovae are quite frequent – on average 20 are seen with telescopes every year. Most of them, however, are too far away to give clues about the violent death of massive stars. Supernovae visible with the naked eye, on the other hand, are very rare. Only four have been seen in the last thousand years – in 1006, 1054, 1572 and 1604 – until Ian Shelton, a young Canadian astronomer visiting the Las Campanas Observatory in Chile, on the night of 23–24 February 1987 saw a new star appear in the Large Magellanic Cloud, a galaxy only 170,000 light-years away.

But SN1987A did not explode in 1987. Although the Large Magellanic Cloud is a close neighbour of the Milky Way, much nearer than other galaxies such as Andromeda, light from it still takes 170,000 years to reach the Earth. In fact, the explosion seen by Shelton that night in 1987 happened during the Ice Age when modern man was evolving from ape-like ancestors.

The progenitor – the star that exploded – was clearly visible on previous surveys and was identified as Sanduleak -69° 202, a blue supergiant. The event, catalogued as SN1987A, gave astrophysicists a unique chance to test their theories about supernovae and a worldwide monitoring campaign was launched, both with ground-based telescopes and satellites. As well as visible light, and ultraviolet and infra-red radiation, SN1987A emitted a characteristic gamma ray sniff of nuclear cooking.

A supernova explosion produces large amounts of the radio-active isotope nickel 56, which decays in turn into the longer-lived cobalt 56 and finally stable iron. The amount of cobalt-56 produced by SN1987A was estimated as 70 times the mass of Jupiter. These processes were clearly visible in SN1987A's "light diary".

Astronomers were surprised by SN1987A. According to supernova theory the progenitor star should have been a red supergiant, not a blue one. It was much hotter than expected (150,000 degrees) and much smaller (only 50 times bigger than the Sun. Astronomers also were puzzled by the fact that a pulsar (see page 110), failed to appear. Others believe SN1987A became a neutron star, but probably within a few minutes some of the exploding material fell back in again and made the star collapse into a black hole.

DOUBLE LIFE

In March 1993 astronomers witnessed a new spectacular supernova, SN1993J, in spiral galaxy M81 in the constellation Ursa Major. The progenitor was a reddish-yellow supergiant, 200 times the diameter of the Sun and with 10 to 18 times the mass. Although 11 million light-years away and not visible to the naked eye, it has provided remarkable information and become a mystery. The supernova seems to have had problems deciding whether to be a so-called Type I supernova, normally occurring in double star system as a white dwarf draws material from a companion star, or a Type II supernova, the classic death of giant stars. It may have ended up being both – a supernova with a double life.

Stellar lighthouses

NEUTRON STARS AND PULSARS

The spent cores of exploding stars collapse into a ball of tightly packed neutrons. These neutron stars spin rapidly, sending out regular radiation pulses which sweep across the sky like the beam from a lighthouse – a "pulsar". These were first seen in 1967, opening a new era in astronomy.

Supernova cores are crushed out of all nuclear recognition, with protons and electrons rammed together to form neutrons. These compact relic neutron stars, predicted by the Russian physicist Lev Landau in 1932, are only about 30 kilometres (20 miles) across but have a density of about a 100 million ton(ne)s per cubic centimetre,

The neutron star surface is made of iron compacted into cylindrical atoms and is very stiff. Beneath this shell are different kinds of fluid. The star's gravity is so strong that a dropped coin would hit the surface with half the speed of light! Only very low, flat beings could ever live on a neutron star; to be tall they would need an extremely strong skeleton!

Although neutron stars shine, they are much too small to be seen directly. However such a star spins rapidly and has a very strong magnetic field. It sets up a powerful particle accelerator, producing a high-energy radiation beam. The beam sweeps across the sky each time the star rotates, setting up a pulsar – a pulsating radio star.

SCRUFF

Jocelyn Bell, a Cambridge research student, working at the Mullard Radio Astronomy Laboratory, discovered the first pulsar – catalogued CP1919 – in 1967. Routinely analyzing 100 feet of paper chart a day looking for new radio sources, she spotted what she called "bit of scruff". Her professor, Anthony Hewish, encouraged her to keep tracking the mysterious signal. For a month nothing happened, but then it started showing up again.

An artist's impression of a pulsar
Electrons trapped in the neutron star's magnetic field are hurled around, accelerated along the field lines. Bending the electrons makes a "screech" of high-powered radiation, which is channelled into narrow beams around the magnetic poles. The pulsar rotates to produce a characteristic "blinking".

The pulses looked suspiciously man-made, coming at an amazingly regular beat, one-and-a-third seconds apart. It almost looked like some distant civilization 200 million light-years away was trying to make contact. Bell and Hewish first nicknamed the signals LGM for "little green men". When announcing their results in February 1968 they were still not sure what kind of source this was. British cosmologist Thomas Gold quickly identified it as a rapidly rotating neutron star. The name "pulsar" was soon coined.

Today more than 600 pulsars are known. They not only produce radio signals, but also other forms of radiation. The pulsar in the famous Crab Nebula sends out both X-rays and gamma rays. Some are incredibly rapid. The fastest known pulsar, PSR1913 +16, is flashing on and off 600 times per second. Pulsars are good timekeepers. Having a "jitter" of only 0.00006 millionths of a second per century they are even more accurate than atomic clocks.

VIBRATO

In 1974, two American astrophysicist, Joseph Taylor and Russell Hulse, found a strange pulsar, PSR1913+16, with a rhythm that had a tiny vibrato. Rather than a single neutron star, this was a pair of stars. The tiny vibrato was produced as the pulsar swung to and fro in its orbit, circling its non-pulsar companion once every eight hours at a distance a little more than the distance between the Earth and the Moon. This compact "binary pulsar" is a very special astronomical physics laboratory, perfect for testing an effect of Einstein's general theory of relativity. Taylor and Hulse were able to show that the two star were approaching each other and speeding up their orbital rotation, like an ice skater's pirouette. The frequency increased by some 76 millionths of a second each year.

The effect is caused by gravitational waves, tiny ripples in space-time produced by the strong gravitation pull between two stars. It was the first time an effect due to gravity waves had been detected, and Taylor and Hulse were awarded the Nobel Prize for Physics in 1993. It was the second pulsar Nobel; Hewish had received the prize for their discovery in 1974.

Neutron stars may not be the only pulsar sources. There are a lot of pulsars in globular clusters, and these can not have been produced in supernovae, otherwise the clusters would have blown apart. These pulsars may be produced by the next most dense kind of star, a white dwarf, which is the closest to the critical mass limit, accretes material from its neighbourhood and implodes.

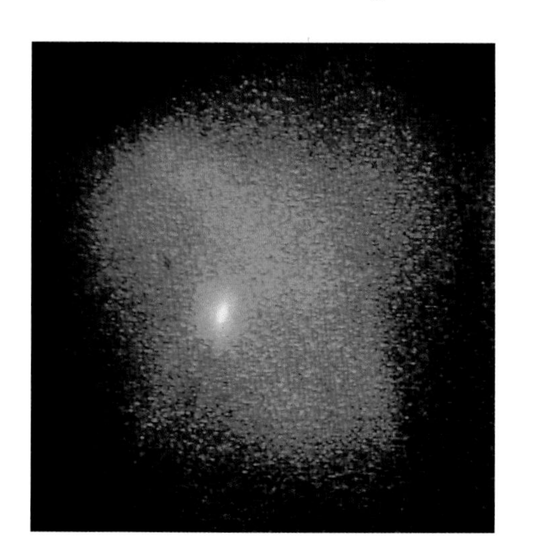

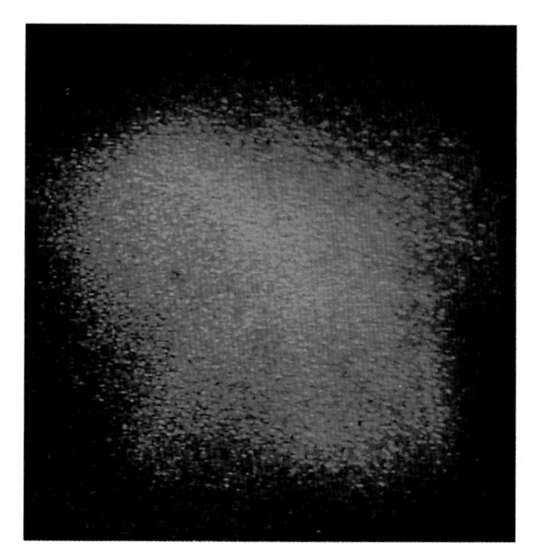

Flashing pulsar *X-ray images of the Crab Pulsar in the Crab Nebula, showing the vivid contrast between the pulsar's "on" (top) and "off" (above) phases. The images were recorded by the Einstein Observatory X-ray astronomy satellite. The Crab pulsar, also known as Taurus A, is blinking 33 times per second.*

JOCELYN BELL – PULSAR LADY

Born in Belfast, Northern Ireland, in 1931, the daughter of an architect, Jocelyn Bell in her early teens was fascinated by astronomy, reading books her father borrowed from the library. She studied physics at Glasgow University, the only girl in a class of 50, then took a Ph.D. at Cambridge in radio astronomy. After her 1967 discovery and subsequent pulsar work, she switched to gamma and X-ray astronomy. Marrying she took the name Bell Burnell. Today she is Professor of Physics in the United Kingdom's largest university, the Open University, and hopes that her position will encourage more women to study physics. She is a practising Quaker and finds it easy to reconcile science and religion. Pulsars are still her "babies" and she keeps pace with research. "Pulsars are a funny subject, now over 25 years old, and I would have guessed that any subject that old begins to get old. But pulsars keep throwing surprises at us", she says.

The dawn of neutrino astronomy

PARTICLES FROM SUPERNOVAE

Several hours before the first light from the big 1987 supernova explosion, bursts of neutrinos were picked up by underground detectors in Japan and the United States. These 19 extragalactic neutrinos from the exploding star announced a new chapter in astronomy.

The gravitational collapse of a supernova releases an immense amount of energy, equivalent to ten million hydrogen bombs exploding simultaneously on each of the 100,000 million stars in our galaxy! Almost all (99.99 per cent) of this energy is produced as neutrinos, formed in the intense temperatures of the collapse.

Although neutrinos streak virtually unhindered through ordinary matter, the inner supernova particles have a hard time fighting their way through the wall of shock enveloping the collapsed star. A few seconds after the gigantic explosion, an outer shell becomes neutrino-transparent, and some particles fly out into space.

With these neutrinos travelling at or near the speed of light, they are difficult or impossible to overtake. Travelling out through space, they herald the supernova, and should have given a warning of the explosion in the Large Magellanic Cloud seen on 23 February 1987.

Although supernovae shine brightly, in fact only a tiny sliver of supernova energy is produced as radiation. When the shock wave from the collapsed core finally blasts away the star's outer shrouding, this radiation is unmasked and the supernova starts to shine. But the neutrinos that were emitted had a head start.

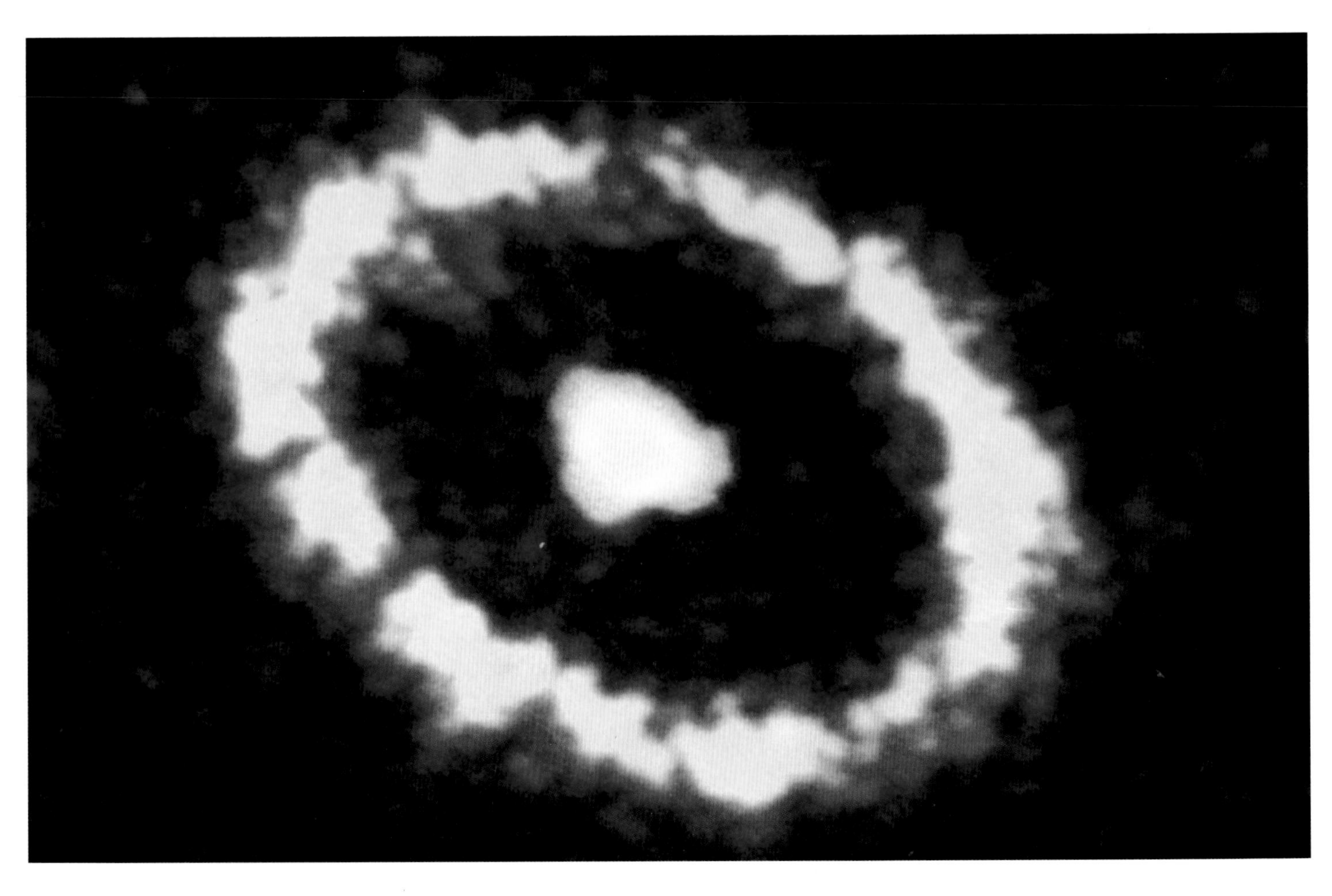

Ringing the changes *A luminous ring of gas around Supernova 1987A seen by the Hubble Space Telescope. The ring, 1.4 light-years across, was blown off the star 10,000 years before it finally exploded. The gas in the ring was heated to fiery incandescence by the huge blast.*

PATIENT WAIT

In 1987 several major experiments were patiently waiting for the tell-tale signs of proton decay (see page 84). These big detectors were also sensitive to extraterrestrial neutrinos. Not expecting startling results, and used to averaging signals over months, if not years, the scientists involved did not usually analyse the data very quickly. Nobody was looking for supernovae. But on hearing the supernova news, and prompted by alert astrophysicists, the experimenters rushed to scan their latest data.

On 10 March, the group operating the Kamioka underground detector in Japan reported a strong neutrino signal from 23 February, three hours before the visual sighting. Some 10^{58} neutrinos were produced as the star collapsed. Of these 300 million million passed through the detector – and just 11 were recorded!

The Kamioka detector is a vast tank filled with water. The high-energy particles from a chance neutrino encounter force their way through the tank faster than the speed of light in water, producing an optical shock wave – Cherenkov radiation (see page 114) – picked up by arrays of light sensors.

Neutrino detector (above) *A technician at the Kamioka neutrino detector in Japan walks through the tank lined with photo tubes before it is filled with water.*

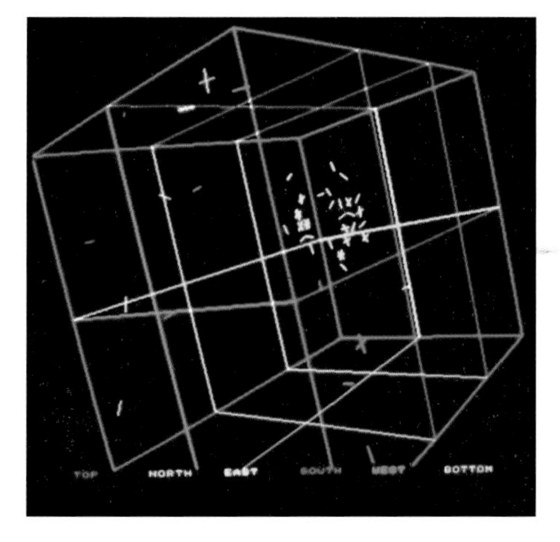

Supernova hit (right) *One of the eight 1987 neutrino "events" as seen by a detector in the Morton-Thiokol salt mine near Painsville, Ohio. In their 7,000 ton(ne) water tank 600 m (2,000 ft) underground a neutrino strikes a proton, producing and neutron and an electron. The yellow crosses mark the light sensors that record electrons.*

LUCKY BREAK

The Kamioka team had been lucky. The magnetic tape on which the Kamioka data is recorded is periodically changed. A routine change would normally have been in progress at the supernova time, and the event would have been missed. However, that day was a Japanese holiday, and the routine tape change was skipped! As if to make up for this stroke of luck, the Kamioka team were penalized by a brief power cut, which had stopped the clock on their detector. Nonetheless, they were still able to fix the supernova signal to within a minute.

One day later, a team of scientists from the University of California at Irvine, the University of Michigan and Brookhaven National Laboratory, operating a water-tank detector in the Morton-Thiokol salt mine near Painsville, Ohio, confirmed the Kamioka result, finding eight supernova neutrino hits themselves.

Astrophysicists were jubilant. The time difference between the visual supernova sighting and the Japanese and American results corresponded exactly to their understanding of supernova formation. Rapid calculations also showed that the neutrino, often assumed to have no mass, has at most a few thousand millionths of the mass of the proton, otherwise the supernova light would have caught up with the advance guard neutrinos during the 160,000-year transit through space. Today, underground detectors routinely scan for signs of supernovae and other celestial fireworks.

Cosmic accelerators

THE SOURCES OF COSMIC RAYS

Astrophysicists have long been puzzled by the origin of cosmic rays, particles with energies millions of times higher than anything manufactured on Earth. Supernova explosions are probably the prime source inside our galaxy, but cosmic rays could also come from distant neutron stars.

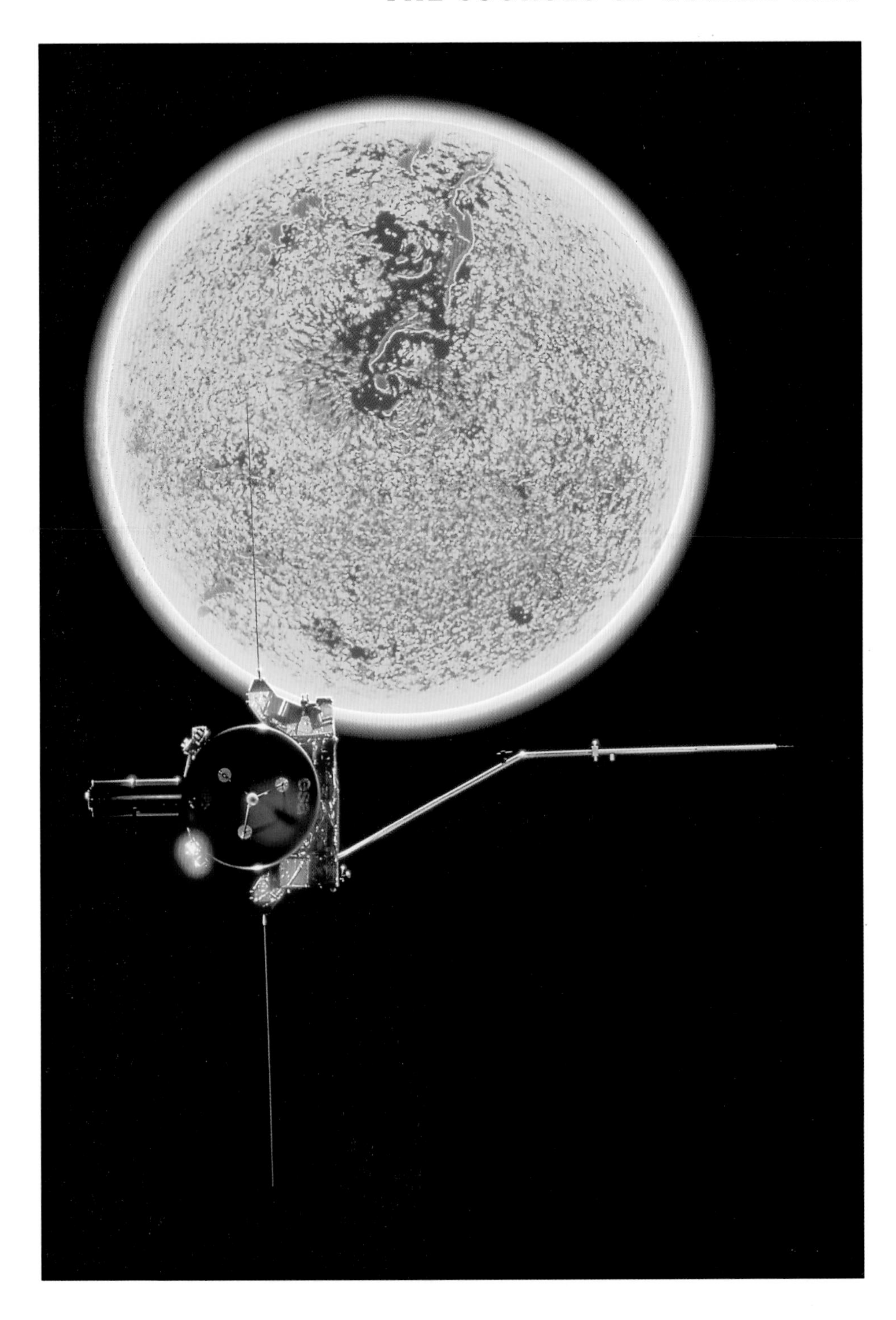

The high-energy particles constantly bombarding the Earth's atmosphere – "primary" cosmic rays – are mainly protons (the nuclei of hydrogen) together with a small fraction of heavier nuclei, mostly helium, and a sprinkling of electrons. Most of these particles probably come from inside our galaxy, which gives out more energy as cosmic rays than it does as radio waves and X-rays combined.

Electrically charged cosmic rays spiral and weave around in the all-pervading magnetic field of the galaxy. Although the galaxy is some 100,000 light-years across, cosmic particles can easily log 20 million light-years before reaching the Earth. Smashing into the gas of the Earth's upper atmosphere, the primary particles produce rich showers of secondaries that rain down on Earth (see page 42).

CHAOTIC PATHS

With their trajectories intertangled, cosmic rays appear to arrive from all directions at similar rates. This chaos makes it hard to see where they come from, and the origin of cosmic rays has always been controversial. Following an initial idea from the famous Italian physicist Enrico Fermi, physicists believed that cosmic rays were emitted by low-energy sources and subsequently accelerated in a series of distinct steps. Successive supernova blast waves kicked the particles to higher and higher energies, eventually attaining velocities close to that of light.

Today most astrophysicists think cosmic rays come from distinct sources that take their particles directly to very high energies. Supernova bursts and supernova remnants – ballooning gas clouds from

PAVEL CHERENKOV'S BLUE BOTTLE

Pavel Alekseivitch Cherenkov was born in a poor peasant home in Voronezh, Russia, in 1904. At Moscow's Lebedev Institute in 1934 he was asked to look at the way radioactivity from radium was absorbed in different fluids, and noticed that a bottle of water exposed to a radium source gave off a faint blue glow. For several years, this mysterious "Cherenkov radiation" was not understood, but it became a major experimental tool for physics. In 1958, Cherenkov and fellow Soviets Ilya Michaelovitch and Igor Tamm received the Nobel Prize for Physics for this discovery.

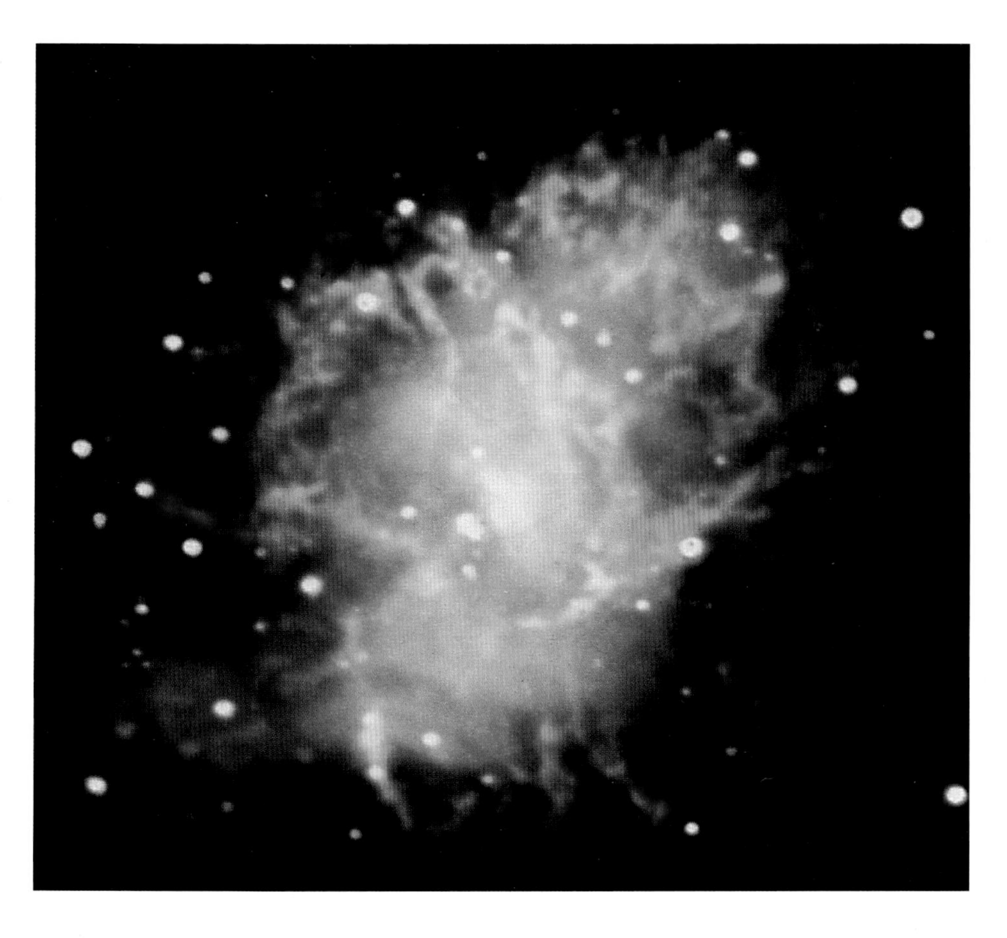

Supernova remnant (above) *The Crab Nebula 6,500 light-years away is the debris of a star that exploded in 1054 and a likely source of cosmic rays. The gas is ballooning out at 1,450 km (900 miles) per second.*

Over the Sun (left) *An artist's impression of the European spacecraft* Ulysses. *It was designed to pass over the Sun's poles in 1994–95. From this vantage point the spacecraft can pick up cosmic rays both in the Solar system and out towards the galaxy.* Ulysses, *built by the European Space Agency (ESA), was launched from the US Space Shuttle* Discovery *in October 1990.*

recently exploded stars – are the most likely lower-energy sources, below 10,000 million electronvolts. Electrons moving close to the speed of light in supernova remnants such as the Crab Nebula could produce great loops of electrons, mirroring the structure of the galaxy.

High-energy particles can also be produced inside our galaxy by a powerful binary system: a neutron star sucking out material from a partner star in orbit. The different rotation speeds of the neutron star and the sucked-out material ("accretion disc") produces a giant cosmic dynamo, accelerating particles to extreme energies. Variations in the rate at which material is sucked out of the companion stars strain these dynamos to the limit, bending the magnetic field lines out of all recognition. The magnetic field eventually snaps back, producing characteristic bursts of radiation and making the source look sporadic.

As well as such cosmic dynamos, neutron stars and pulsars could also join in the act. A contender is Cygnus X-3, a double star in a spiral arm at the edge of our galaxy, some 37,000 light-years away. Discovered as an X-ray source in the 1960s, Cygnus X-3 became a centre of attraction in 1972 with an amazing burst of radio waves.

BLUE SHOWER

Free-ranging gamma rays are generated by cosmic ray proton interactions. Crashing into Earth's atmosphere, they produce showers of electron–positron pairs. Moving faster than light through the air (light travels more slowly through air than through a vacuum), these particles set up a shock wave of bluish light, like the boom of a supersonic aircraft. The light, discovered by the Soviet physicist Pavel Cherenkov in 1934 and called "Cherenkov radiation", can be picked up by ground-based detectors.

The origin of cosmic rays is a deepening mystery, especially ultra-high-energy cosmic rays, with energies of more than 10^{20} electronvolts – millions of times greater than those of the biggest particle accelerators. These particles may be immigrants from outside our galaxy, blasted in by huge intergalactic cyclones generated by the steady accumulation of energy from many supernovae.

Ultimate collapse

ENTER THE BLACK HOLE

Large stars are eventually doomed to total gravitational collapse. After their grand supernova finale they are too big to end as white dwarves or even neutron stars, but are condemned to disappear into a "black hole" where gravity is so strong that it swallows everything. Not even light can escape.

If a rocket exceeds the Earth's critical "escape velocity" (see page 102) of 11 kilometres (7 miles) per second, it eludes gravity and sails out into space. At lower speeds it will fall back like a stone, or go into orbit, circling the globe as a satellite. This escape velocity depends on the gravitational mass. To escape from the Sun needs a velocity of 620 kilometres (385 miles) per second, while a neutron-star exit permit requires 200,000 kilometres (125,000 miles) per second.

In 1783, the English astronomer John Michell realized that if a star were heavy enough, its escape velocity would exceed 300,000 kilometres (185,000 miles) per second – the speed of light. Light would not be able to escape and the star would be invisible. Building on Newton's theory that light is particles, he believed that gravity would directly affect the light from a star and try to pull it back. "All light emitted from such a body would be made to return to it by its own power of gravity", he wrote.

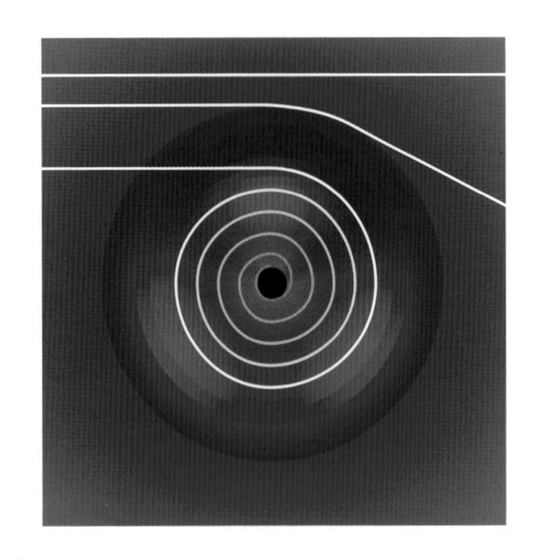

Michell's physics was wrong – the speed of light is unaltered by gravity – but he arrived at the right conclusion.

Magic circle *The event horizon, the border of a black hole, lies at the Schwarzschild radius, named after the German astrophysicist Karl Schwarzschild who in in December 1915 used Einstein's then new equations of general relativity to calculate the gravity around a sphere. The radius draws a sphere beyond which no light can escape. Light rays approaching a black hole will be bent and spiral into its deep gravity well. The radius depends on the collapsed star's mass and is relatively small. For a star ten times the mass of the Sun, the radius will be just 30 km (20 miles). "There is a magic circle which no measurements can bring us inside", said Arthur Eddington.*

In Einstein's general relativity, light faithfully follows the curvature of space around heavy bodies such as stars. A collapsing star several times heavier than the Sun creates a gravitational "well" from which light has more and more difficulty escaping. Eventually the light is completely trapped and the star becomes a "black hole".

The possibility of such a total gravitational collapse was pointed out by Robert Oppenheimer in the United States in 1939. However, he thought it was just a quirk of the equations of relativity and had no real meaning, and left to become the senior scientist in the atomic bomb project. Apart from a small band of black hole enthusiasts, the idea lay forgotten until the early 1960s when new sky surveys revealed intriguingly powerful gravitational sources in the depths of space.

The American theoretician John Wheeler coined the term "black hole" in 1969, after a revival of interest in these stellar catastrophes. "A black hole has no hair", Wheeler once said, meaning that nothing whatever emerges from a black hole.

Swan song ***An artist's impression of Cygnus X-1 in the constellation Cygnus, the Swan, as if seen from a nearby asteroid. Cygnus X-1 revolves around a hot blue star, sucking matter from it, which swirls around like water going down a drain.***

During this time, Roger Penrose and Stephen Hawking at Oxford showed that a black hole contained a "singularity" of relativity – a zero point of infinite density where physics collapses, imploding on itself and making predictions impossible. Hawking became intrigued by these singularities. The Universe itself was created in a time-reversed version of such a catastrophe – the Big Bang – when physics exploded into existence.

EVENT HORIZON

A black hole is hidden out of sight, lurking inside its "event horizon" – the surrounding sphere where space is so curved that light, and everything else, cannot escape. Everything that has been sucked into the black hole lies hidden inside. The event horizon also screens permanently from view the physics singularity at the hole's centre.

A clock falling into a black hole would appear to run slower and slower, gradually becoming redder and fainter and finally disappearing from view. Extended objects, such as astronauts, would be torn apart on approaching a black hole, the gravitational pull at the end nearer the hole being much stronger than at the other end.

FAVOURITE CANDIDATE

Stars with about ten times the mass of the Sun or more are candidates for black holes. Although an isolated hole is invisible, a black hole often rotates in a bizarre *pas de deux* with a nearby star. Several possible black holes have been observed. One of the favourite contenders is Cygnus X-1, 6,500 light-years away. As it sucks matter from an orbiting companion star, X-rays are generated as matter spirals into the black hole. Black holes may lurk at the heart of most galaxies, marking a common grave for many ancestor stars.

Black-hole power

QUANTUM EFFECTS IN THE COSMOS

Black holes called for imaginative physics. They can radiate, so are not really black. Revolutionary black hole thinking brought together the two big theories of the twentieth century – general relativity and quantum theory – which had previously refused each other's company.

Black holes spell trouble. Their arrival on the physics scene in the 1960s called for a radical rethink of several time-honoured ideas. For example when matter disappears into a black hole, less matter is necessarily left behind. Black holes could operate as a kind of cosmic vacuum cleaner, keeping the the Universe neat and tidy. But this violates a sacred physics law that says that, left to itself, the Universe prefers disorder – called "entropy".

This paradox was resolved when the fertile mind of Stephen Hawking probed deeply into black hole ideas in the early 1970s. While nobody can get inside information on a black hole, the event horizon, the frontier where light becomes trapped, does give some clues to the hole's appetite. When matter is sucked in, the hole gets heavier and its event horizon gets bigger. Hawking's new ideas suggested to Jacob Bekenstein at Princeton that the event horizon is a measure of the invisible disorder lurking inside.

Then came another problem. Entropy is intimately linked with temperature. If a black hole had entropy, it should also have a temperature. But to have a temperature, a body must radiate. Even black holes had to emit something. On a trip to Moscow in 1973, Hawking was convinced in discussions with Soviet cosmologists that the conundrum could be resolved with the wizardry of quantum theory.

According to the uncertainty principle (see page 36), even a total vacuum is not empty, but full of quantum fireworks powered by "borrowed" energy. In quantum terms, energy is lent free of charge as long as it is paid back quickly enough – before Nature has time to notice.

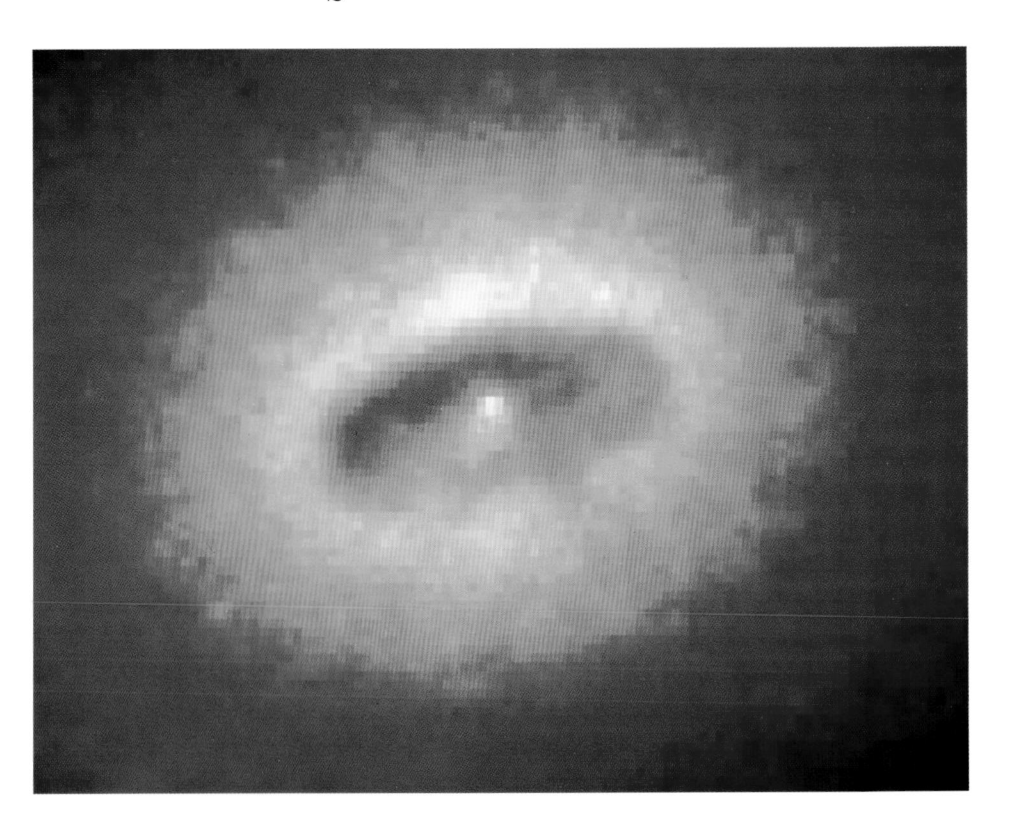

HOLES THAT EXPLODE

When such a quantum blip happens near a black hole, the energy bookkeeping is affected by the huge gravitational force. If both particles fall into the black hole, nobody is any wiser. But if only one particle falls in, the black hole can absorb the energy debt and the other particle is suddenly free. To someone watching the black hole from afar, it looks as though the hole has radiated a particle.

Swallowing the energy debt reduces the mass of the black hole, according to Einstein's $E=mc^2$, so a black hole is continually "evaporating" – getting smaller and hotter. However, the rate of evaporation of normal black holes, formed by the collapse of stars, is negligible. With a temperature of less than a millionth of a degree above absolute zero, radiation is practically non-existent.

In 1971, early in the black hole game, Hawking boldly suggested that in the immediate aftermath of the Big Bang, isolated concentrations of temperature and pressure could have formed much smaller black holes, as small as 10^{-13} centimetres (inches) across, about the size of a proton, but still weighing many millions of ton(ne)s. Hawking's calculations also showed that black hole temperature is inversely proportional to the mass – the smaller the black hole, the higher its temperature and the more it radiates. Small black holes should therefore be easier to see than big ones!

Seeing a black hole (left) *A Hubble Space Telescope image at visible wavelengths of a giant disc of hot gas and dust fuelling a suspected black hole. The disc, estimated to be 400 light-years across, surrounds the core of the active galaxy NGC 4261 in the Virgo cluster, some 45 light-years away. The bright central core is thought to contain a black hole, sucking in energy from all around.*

Evaporating black hole (below) *Quantum uncertainty allows pairs of particles and antiparticles to pop out of empty space right outside the border of a black hole. One member of a pair may fall into the hole, while the other escapes ("Hawking radiation"). As black holes emit particles in this way they lose mass and size, eventually disappearing.*

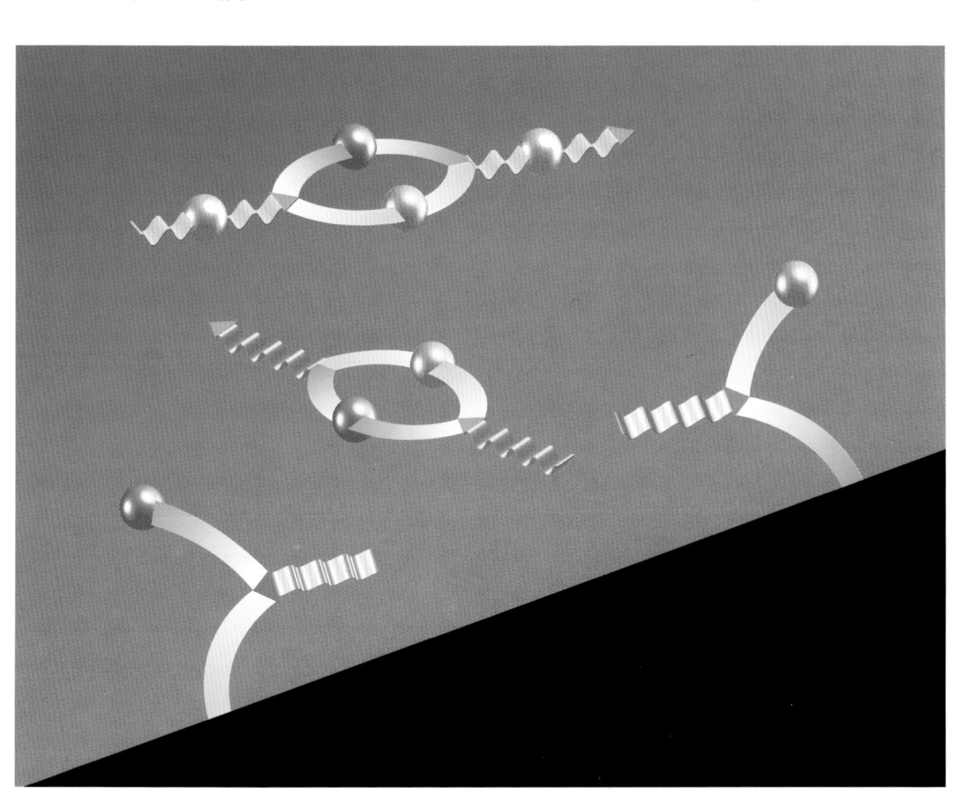

Finally a microscopic hole will come to an end in a massive explosion. Many of these primordial "mini black holes" have probably already evaporated, disappearing in a gigantic shower of gamma rays; others are nearing the ends of their lives and could soon die in a crescendo of radiation. Experiments have looked for these flashes, but no convincing signal has yet been seen.

COMBINED THEORIES

Stephen Hawking's prediction of black hole radiation, now called Hawking radiation, was the first success in combining Einstein's theory of general relativity with quantum theory, two cornerstones of twentieth century physics. As often happens when radical new scientific ideas are presented, the proposal was initially greeted with incredulity when Hawking launched it in 1974 at the Rutherford Laboratory near Oxford.

This work blazed a new trail for attempts to bring the two cornerstone theories together. Each theory works well in its own domain of physics, general relativity describing the large-scale Universe and quantum theory the subatomic world, but harnessing them together presents enormous difficulties. This would pave the way for an all-encompassing "theory of everything".

HAWKING AND PENROSE – PROVING THE BIG BANG

In 1965, Roger Penrose in London showed that a star collapsing under its own gravity eventually creates a "singularity" – a point of infinite density where Einstein's general relativity and its laws of space and time break down. At such a singularity, no calculations can be made, and therefore no predictions are possible.

Penrose and Stephen Hawking (who was at that time a research student at Cambridge) then turned the process round. Just as a collapsing star ends in a singularity, they showed that an expanding Universe must have started with one.

Hawking draws an analogy between these cosmic cataclysms and the inability of classical physics to describe stable atoms (see page 34). Only when quantum physics was developed in the 1920s could physicists understand how the atom was held together. In a similar way, suggests Hawking, quantum theory could rescue general relativity from these incalculable singularities.

A big new eye in the sky

THE HUBBLE SPACE TELESCOPE

A new era in astronomy opened in 1990 when the Hubble Space Telescope went into orbit around the Earth. For the first time a large telescope was not obscured by the Earth's atmosphere. Having been repaired by astronauts in December 1993, the telescope now is sending back marvellous images.

Astronomers have long dreamt about an observatory above the atmosphere which would give them an unobscured view of the Universe. The atmosphere distorts light from stars and galaxies, making the stars twinkle in the sky. Even telescopes built on high mountain tops to minimize the effect only collect a blurred image. Looking at stars from Earth is rather like bird-watching from the bottom of a swimming pool.

The astronomer's dream finally came true when the Hubble Space Telescope (HST) was deployed from the Space Shuttle on 24 April 1990. Named after the great US astronomer Edwin Hubble, the 11-ton(ne) satellite was designed to explore the heavens in ten times more detail than before. The telescope initially had blurred vision due to a faulty main mirror and also had a mechanical problem with "shaking" solar panels.

Sophisticated image processing and spacecraft control compensated for these major problems, and so HST during its first three years in orbit catalogued a rich harvest of observations, including compelling clues to the existence of several super-massive black holes (see page 128) and amazing close-ups of the Orion nebula, where new stars are born from clouds of gas. A vast cloud of gas in Orion known as Herbig-Haro No.2 in June 1993 was seen heated by shock waves from jets of high speed gas ejected from a new-born star. This splendid shot give astronomers an unprecedented glimpse of star formation.

Hubble also beamed back detailed images of the supernova remnant Cygnus Loop, known as the Veil Nebula. The telescope also sighted a new type of cosmic object – a gigantic concentration of stars produced by two colliding galaxies 200 million light-years away , called a "starburst galaxy". HST in February 1992 saw a ballooning shell of gas, 400 times the diameter of our solar system, coming from a nova (Cygni 1992) – a thermonuclear explosion on the surface of a white dwarf star.

One of the most spectacular Hubble pictures shows the famous "Einstein Cross". As light from a quasar 8,000 million light-years away grazes a galaxy at only 400 million light-years, it is bent in new directions. From Earth we see four images of the distant quasar with the foreground galaxy in the middle. The effect, called "gravitational lensing", was predicted by Albert Einstein. Light can be bent by massive objects like stars and galaxies, he said. This gravitational lensing provides a new method of detecting otherwise invisible stars (see page 132) as they cross in front of brighter ones.

New eye in space (above) *The Hubble Space Telescope being deployed in orbit 610 kilometres (380 miles) above the Earth.*

Hubble Space Telescope images (right-hand page) *The unparalleled depth of HST images opens of a new astronomical wonderland: the Einstein Cross (top left), a distant quasar seen through a gravity lens; the spiral galaxy M100 (middle left); the core of active galaxy NGC 1068 (bottom left); the supernova remnant Cygnus Loop (right), 2,500 light-years from Earth, is the remnant of a star that exploded 15,000 years ago at the time of the Cro-Magnon Man – the picture reveals a network of gas heated to 60,000 degrees by shock waves from the explosion.*

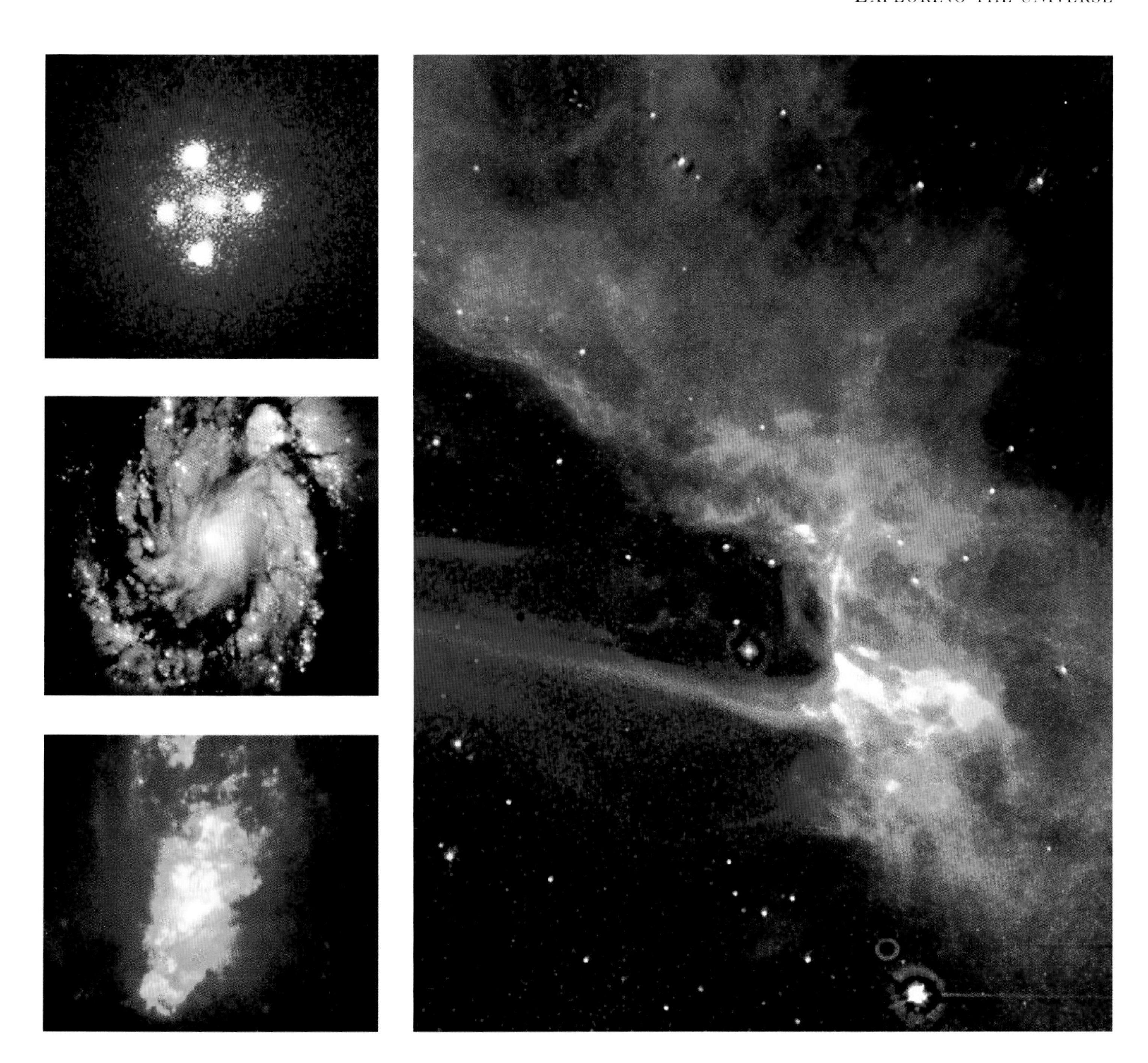

A GIANT LEAP
In December 1993 intrepid astronauts from the Space Shuttle *Endeavour* equipped HST with optics correcting the blurred vision and new solar panels. "A small change for a mirror, a giant leap for astronomy", Christopher J. Burrows of the Space Telescope Science Institute said when the first new Hubble images were presented to the public in January 1994.

The mighty spiral galaxy M100 which lies several tens of millions of light-years away was seen as clearly and with the same detail as was previously only possible for the few nearby galaxies in our Local Group. The Hubble Space Telescope also revealed the central region of an active galaxy, NGC 1068. It lies at a distance of 60 million light-years and belongs to a class of galaxies called Seyfert Type 2. The core shines a billion times more brightly than the Sun. The most likely source of the galaxy's fantastic energy output is a supermassive black hole equalling a 100 million solar masses.

With its improved vision HST will look further into space and explore the wonders of the Universe as they have never been seen before. Astronomers say that the telescope is even better than it was planned to be and in 1994 it showed its new potential. With its unrivalled ability to measure cosmic distances, it could help to answer one of the biggest questions of all: how big, and how old, is the Universe?

The ultra-violent sky

THE UNIVERSE IN A DIFFERENT LIGHT

Visible light is only a small slit in the broad spectrum of electromagnetic radiation. Beyond the vision of even the best optical telescopes is a magnificent invisible Universe. Space astronomy with detectors lofted above the Earth's atmosphere is beginning to reveal a much more violent cosmos.

The era of invisible astronomy began in 1931 when Karl Jansky, working on radio communications at Bell Laboratories in the United States, detected a strange noise coming from the centre of the Milky Way. But his report, "Electrical Disturbances of Extraterrestrial Origin", went unnoticed. After World War II, radio astronomy progressed rapidly, profiting from the advances of wartime radar technology. The radio sky opened a new window on the Universe, and major discoveries followed, such as quasars (see page 128), and the microwave background radiation.

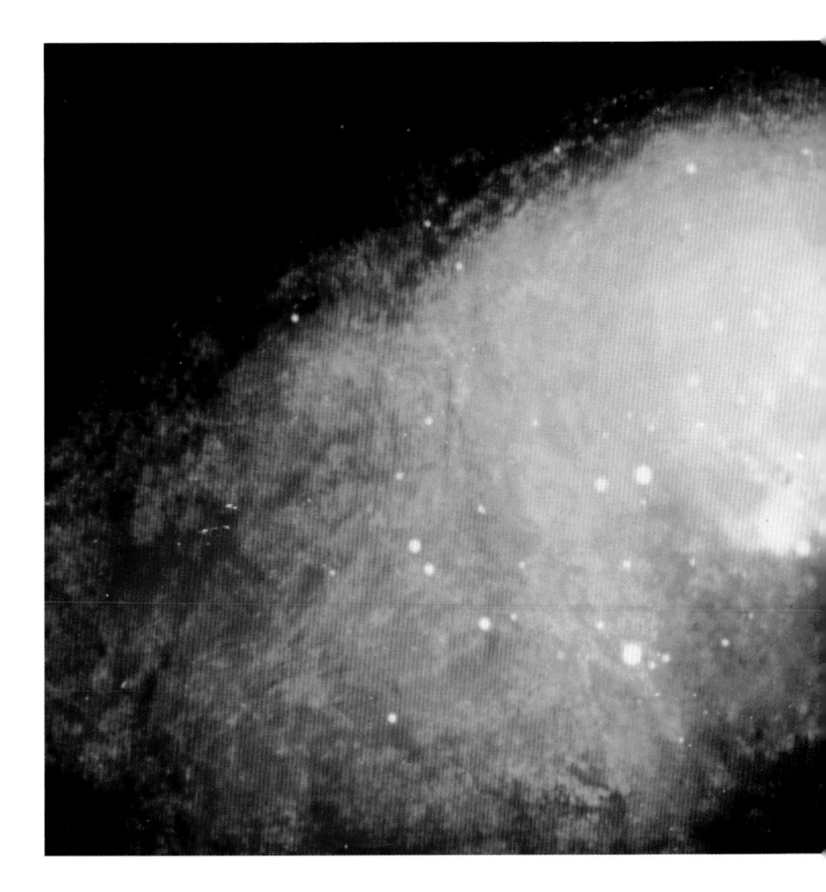

THE HOT UNIVERSE

The success of radio astronomy encouraged astronomers to search other regions of the spectrum. First they turned to infra-red radiation, beyond the red end of light's band of colours. Infra-red rays from space are absorbed by the atmosphere, but can be detected on high mountains.

Infra-red radiation is often called "heat radiation" because our bodies feel these wavelengths as heat; in cosmic terms infra-red radiation is cold, emitted by semi-frigid bodies below 6,000 degrees.

Beyond the other end of the visible spectrum is ultraviolet radiation. This part of the spectrum, successfully observed by the International Ultraviolet Explorer (IUE) launched in 1978, is the threshold of the truly hot Universe, with temperatures of millions of degrees giving radiation extending to X-rays and gamma rays.

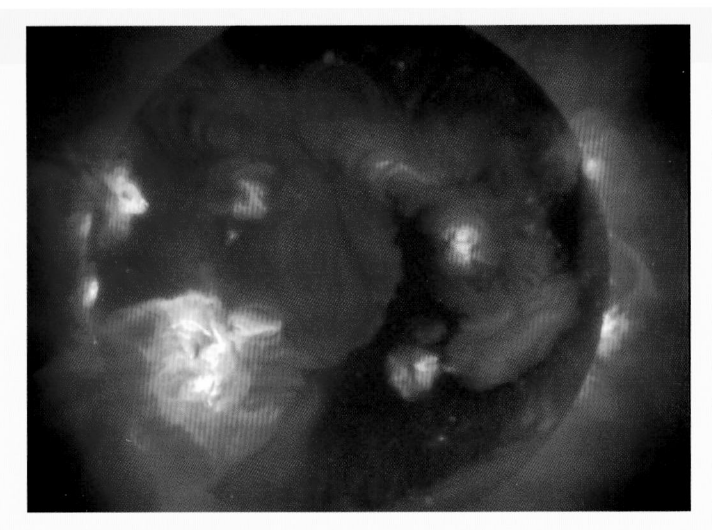

X-RAY SUN

The Sun only gives off a small portion of its energy as X-rays. These comes from the Sun's outer atmosphere, the corona, which is heated to a million degrees by violent upheavals of the solar surface called "flares". Electrons blasting through the thin gas give off a high-pitched X-ray screech as they swerve past atomic nuclei.

Between October 1991 and January 1992, the Japanese X-ray satellite Yohkoh ("Sunbeam"), no bigger than an office desk, recorded a dramatic ten-minute "movie" of the Sun's corona. This is in constant flux, alternately brightening, fading and reforming like a wildfire. Splendid loops of gas cascade out and mighty eruptions at the fringes of the solar surface grow into helmet-shaped structures.

NEW OBSERVATORIES

Three new observatories exploring the Universe in invisible light are planned for the coming years. NASA expects an advanced X-ray observatory (AXAF) to go up in 1998, followed some years later by a new infra-red telescope (SIRTF). ESA, the European space agency, will launch INTEGRAL, a gamma-ray observatory, in 2001.

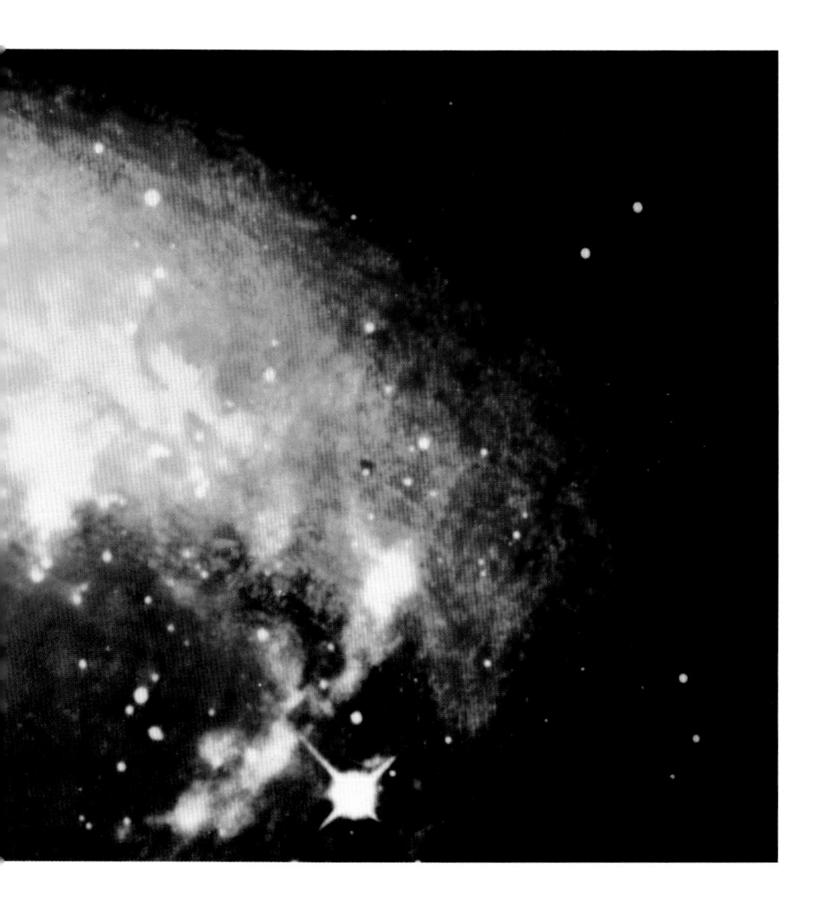

X-RAY SURPRISE

X-ray astronomy dates back to 1948, when American scientists launched a specially equipped German V-2 rocket and discovered that the Sun shines X-rays (see box bottom left). In 1962 a US rocket-borne detector, designed to see radiation from high-energy solar particles crashing onto the Moon, unexpectedly saw the first distant X-ray star, Scorpius X-1. At that time many astronomers were reluctant to believe that stars could shine X-rays.

In the 1960s several new X-ray objects were reported and, with excitement growing over this new astronomy, in 1970 NASA launched the first X-ray satellite, Uhuru (Swahili for "freedom"). By the end of that decade more than a thousand X-ray sources had been catalogued.

RÖNTGEN IN THE SKY

More recently, the German ROSAT (Röntgen Satellite, named after Wilhelm Konrad Röntgen, the discoverer of X-rays) has completed the first ever X-ray map of the entire sky and has catalogued more than 60,000 X-ray sources, including dozens of hitherto unseen supernova remnants. The 2.5-ton(ne) satellite, which was launched in 1990 and is able to discern objects that appear a thousand times smaller than the full moon, has looked deep into nearby galaxies – such as Andromeda – and beyond, seeing distant X-ray quasars 10,000 million miles away and clouds of dark interstellar matter.

Recently ROSAT has provided a fresh view of a familiar star cluster, the Pleiades (Seven Sisters), 410 light-years away and containing some 500 stars. The cluster X-ray view looks very different from the optical picture. These relatively young stars give out large amounts of X-rays, and the six or seven bright stars (the "sisters") easily seen with the naked eye disappear, while otherwise invisible stars dominate the X-ray picture. These are younger stars which spin faster and therefore generate a much larger quantity of X-rays.

Stellar nursery (above) *An infra-red image of the Orion Nebula, a famous region 1,500 light-years away where new stars are born from clouds of condensing gas. Scientists used data obtained from a NASA airborne telescope and superimposed the infra-red image on a black-and-white photograph of the nebula. The colour coding indicate temperatures, red being the coldest regions. At the centre of the image are a group of four close-set stars called Trapezium which are part of a cluster of hot young stars, maybe only 20,000 years old. Slightly to the upper left of the four stars is the present stellar nursery in Orion, invisible to optical telescopes.*

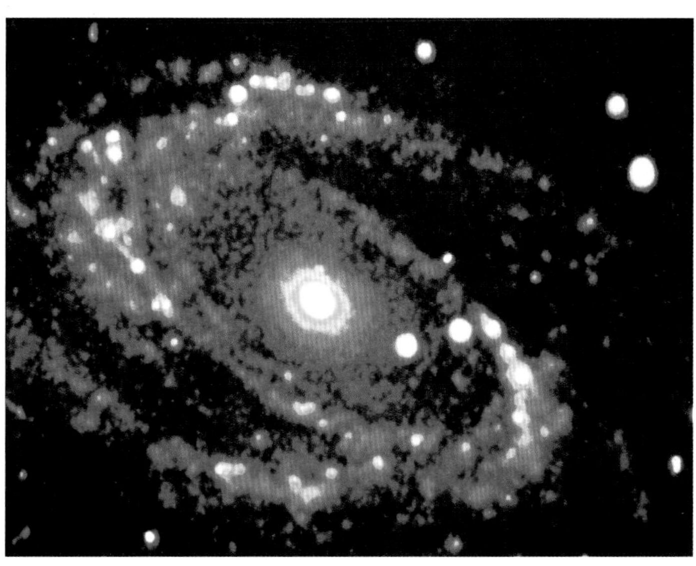

Great Bear (right) *A false-colour image of the M81 spiral galaxy taken by the Ultraviolet Imaging Telescope, part of the* ASTRO-1 *mission of Space Shuttle* Columbia *in 1990. It shows regions of star formation and other structures in the spiral arms. M81 is in the constellation Ursa Major (Great Bear), 10 million light-years away.*

Celestial mystery bursts

THE GAMMA RAY ENIGMA

A new astronomy puzzle is the continual bursts of celestial gamma rays, seemingly random and occurring almost daily. The bursts, now explored by NASA's Compton Gamma Ray Observatory, could come from just outside our solar system or from the far corners of the cosmos.

Gamma rays are produced in a wide range of cosmic processes. These include the supernova explosions, particle acceleration by strong fields, matter-antimatter annihilation and radiation from exotica like neutron stars, black holes, active galactic nuclei and quasars.

The most intriguing phenomena in this part of the electromagnetic spectrum is the mysterious gamma ray bursts, sudden flashes of high-energy gamma radiation lasting from one hundredth of a second to 1,000 seconds. The brief and bright flashes occur at unpredictable times all over the sky and were first seen in 1967 by the US spy satellite Vela, deployed to monitor for the tell-tale gamma bursts produced by terrestrial nuclear explosions.

A search for the mystery bursts was high up on the agenda for NASA's Compton Gamma Ray Observatory, deployed by the Space Shuttle *Atlantis* in April 1991. The 16-ton(ne) satellite, named after physicist Arthur Holly Compton, has four instruments and scans the whole range of cosmic gamma energies. It can pick up signals 50 times fainter than anything observed previously.

The first two years in orbit the Compton Observatory detected on average one gamma-ray burst a day. The most powerful of these was in the constellation Virgo on 31 January 1993, the day of the Superbowl American football game and became

Gamma eyes (right) ***The Compton Observatory being released from the Space Shuttle*** **Atlantis** ***400 km (250 miles) above Earth. Covering a whole range of gamma-ray energies, the satellite's detailed images provide a new window on the Universe.***

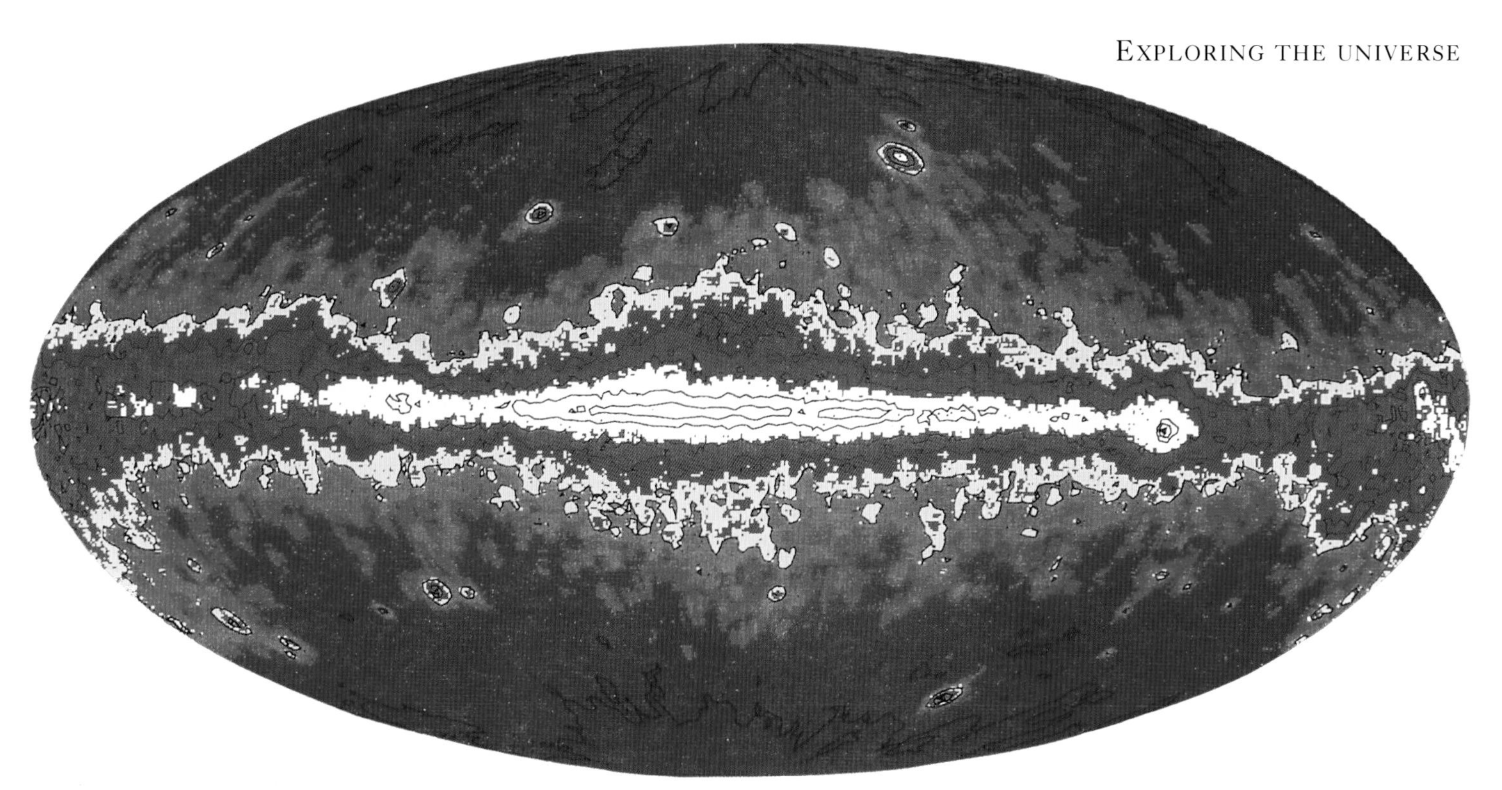

known as the "Superbowl Burst". A hundred times brighter than any steady gamma-ray source inside our galaxy, it temporarily blinded one of the Observatory's instruments.

Right from the start, the source of the gamma ray bursts have been a mystery. Before the Compton Observatory was launched most astrophysicists believed the bursts were caused by seismic explosions or impacts of asteroids on the surface of neutron stars. The observatory has added to the mystery, showing that the burst sources seem to be confined in space and form a spherical grouping with the Earth at the centre.

Scientists have tried hard to find out what celestial objects can cause such a distribution. Some believe that the bursts come from neutron stars that form a huge halo around our galaxy. Other theorists say they come from collisions of comets or other events just outside our solar system – or they could, on the contrary, come from far corners of the Universe.

GEMINGA

The most vivid gamma-ray source in the sky has been Geminga, so named because it lies in the constellation Gemini. The source was discovered in 1973 by NASA's Small Astronomy Satellite (SAS-2) and the mystery of Geminga now seems to have been solved. Compton has catalogued Geminga as a gamma-ray pulsar (see page 110).

Geminga is a rapidly spinning neutron star, the remnant of a giant sun that became a supernova some 340,000 years ago. The pulsar moves through space and astronomers believe the star that exploded was in the constellation Orion, maybe only 100 light-years away. The supernova, maybe 20 times brighter than the full moon and visible in daylight for two years, must have scared our ancestors. The shock wave may have carved out the so-called "Local Bubble", a huge, low-density region around our Sun.

Gamma-ray sky (above) *The first map of the gamma-ray sky as surveyed by the Compton Gamma Ray Observatory. The brightest regions with the most powerful sources are coloured white, the faintest are blue. The horizontal band is the Milky Way galaxy. Geminga is the white spot at the far right of the image.*

Geminga (below) *Compton image of the once mysterious gamma-ray pulsar Geminga (pink spot) and the Crab nebula pulsar (centre). Of the 500 known radio pulsars, only a few are gamma sources. Geminga could be an example of a new type of pulsar that does not emit radio waves.*

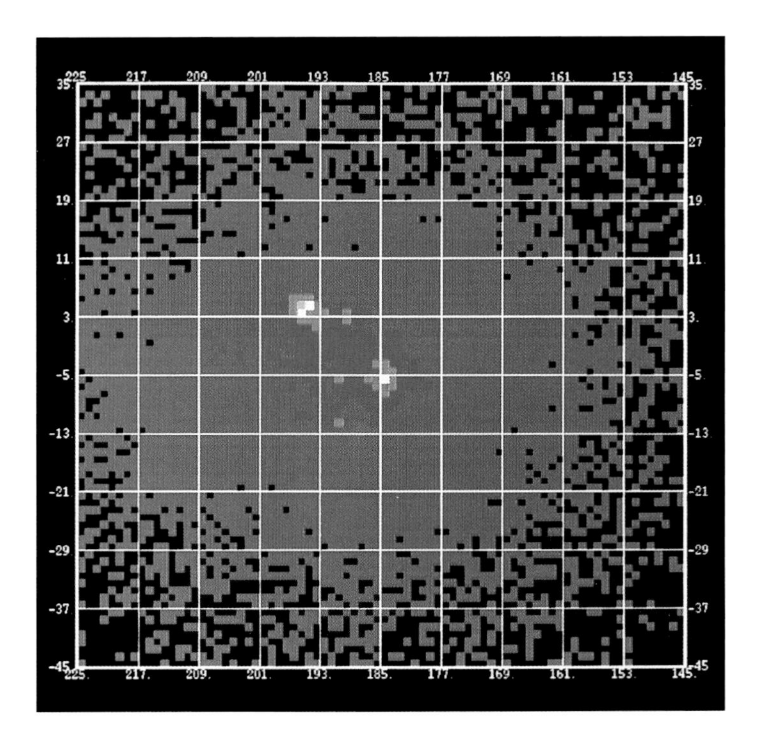

Patterns of infinity

THE LARGEST STRUCTURES OF ALL

While we have discovered smaller and smaller constituents of matter, an intriguing new picture of the large-scale Universe has also emerged. In the 1980s, astronomers found that the galaxies are grouped into larger and more complex structures at greater depths in space.

MARGARET GELLER – FINDING THE GREAT WALL

Margaret Geller of the Harvard-Smithsonian Center for Astrophysics has played a leading role in mapping the furthest reaches of the Universe.

In 1985, with John Huchra, she began a careful survey of 15,000 galaxies out to a distance of 600 million light-years, pushing back the charted regions of the Universe. Carefully covering the sky, strip by strip, Geller and Huchra found that galaxies lie in thin sheets wrapped around vast holes almost devoid of galaxies. The result "looked like a kitchen sink full of soap suds", she commented. This work also revealed the Great Wall, a vast slab of galaxies 500 million light-years across and millions of light-years thick extending right across the north galactic hemisphere.

Before embarking on a career in astronomy, Margaret Geller had aspirations to become a designer. Her work as a cosmic cartographer combines both of these flairs.

People have always looked for patterns in the sky. Using much imagination, the Greeks organized the stars into 48 pictures or constellations with motifs from mythology. They called this "astronomy" (star arranging) and we still use this folklore as a guide to the 100,000 million stars in our Milky Way galaxy.

On a larger scale, galaxies tend to attract each other by gravity, tens or hundreds of thousands of them grouping into roughly spherical "clusters" with a typical diameter of 10 or 20 million light-years. The Milky Way is in a small cluster of about 20 galaxies known as the Local Group, which includes the nearest galaxy to our own, Andromeda.

SUPERCLUSTERS

The nearest major (or "rich") cluster to our own is Virgo, 40 million light-years away, containing a few thousand galaxies. This cluster is itself the centre of the Local Supercluster, which contains at least eleven clusters and 40 additional groups – including our Local Group – and has a total of 50,000 galaxies. We are at the fringes of this giant lens-shaped structure, 100 million light-years across.

At first astronomers, guided by two-dimensional star maps, thought that clusters were spread uniformly across space. The positions of many stars can be measured from the same photograph, but three-dimensional maps are more difficult, as the distance of each galaxy has to be measured separately. A shift of colours in each galaxy's spectrum tells how fast it is moving (see page 96). In the continually expanding Universe, the further away an object is, the faster it recedes, so the speed of a galaxy also indicates its distance.

The Milky Way does not follow the rest of the expanding Universe exactly. We are sliding towards the constellation Centaurus at a relentless 600 kilometres (350 miles) per second, as if tugged by a huge, distant mass, known as the "Great Attractor".

Embarking on new surveys in the early 1980s, astronomers were astonished to find that some superclusters look like sheets, while others are long chains, like pearls on a string. (These string-like structures are probably sheets seen from the side.) Between the superclusters are vast voids containing almost no galaxies.

HOLE IN THE SKY

The biggest cosmic void, found in the constellation Boötes, is 300 million light-years across and surrounded by "walls" of superclusters. This yawning gap, discovered in 1978, was first thought to be the only major hole in the Universe.

What lies ever further out? It may be that superclusters are linked by faint bridges into "super-superclusters". Despite looking deep into the sky, astronomers have so far mapped only a small percentage of the Universe. Each time a sky survey has embarked on a larger-scale search, bigger structures have been found.

Bright cluster (above) *M13 in Hercules, the brightest globular cluster in the northern skies, contains some 500,000 stars and lies 22,500 light-years away. Globular clusters form halos around galaxies and are among the oldest known objects, at least 13 billion years old. Some 125 clusters are known in our galaxy. Clusters are not uniformly spread around the sky, but are themselves grouped into irregular "superclusters".*

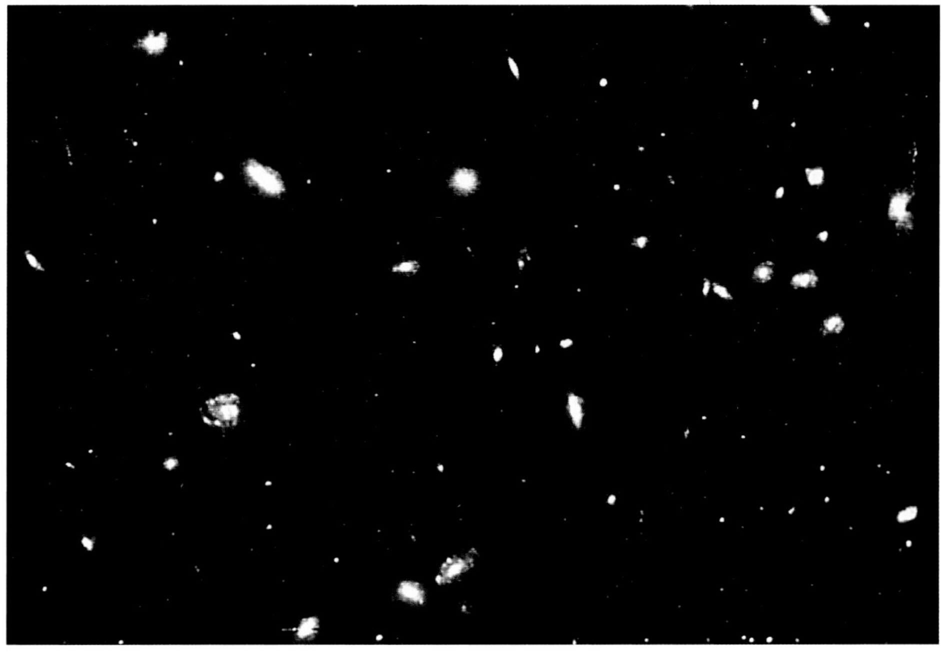

Snapshot of galactic evolution (right) *A Hubble Space Telescope picture of the cluster of galaxies CL 0939+4713 (4 billion light-years away). The cluster contains several times more spiral galaxies than other, more evolved, clusters. The image suggests that some of the spirals are about to collide and merge to form elliptical galaxies.*

Blazing embers

ACTIVE GALAXIES

In the 1960s, radio astronomers found powerful sources that did not correlate with any known stars. At the fringe of the Universe, these quasars blast out incredible amounts of energy. Probably powered by supermassive black holes, they may be the first stages of galaxy formation.

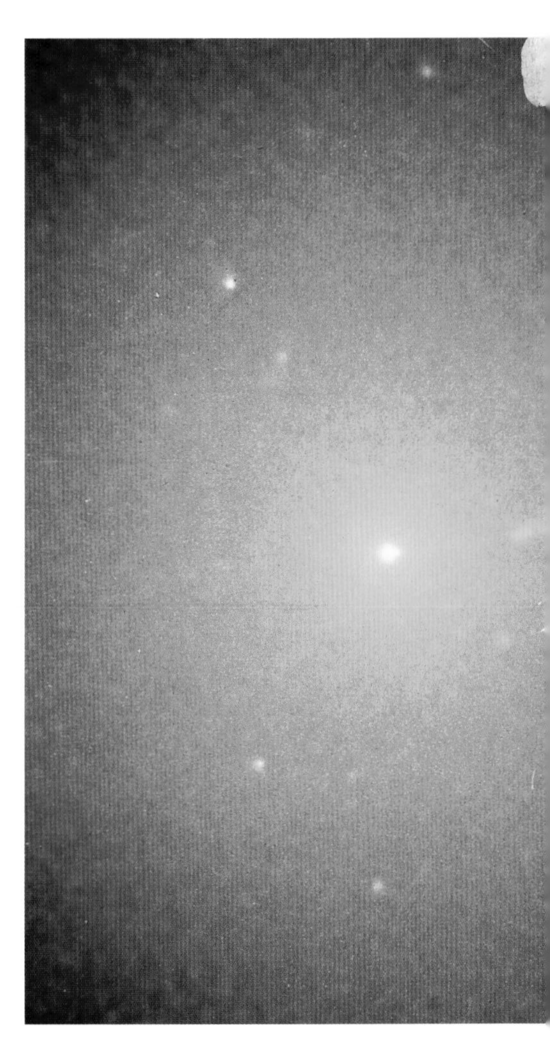

Galaxies are not as tranquil as they first appear when viewed in optical telescopes. In 1943 the US astronomer Carl Seyfert found spirals – now called "Seyfert galaxies" – with very bright centres. In the 1950s, new technology opened up the science of radio astronomy, tracking faint radio signals from the distant sky. Some sources could not be paired with a visible partner, even using the most powerful optical telescopes.

Around 1960, astronomers located sharp radio sources that coincided with what looked like dim stars, so dim that they needed many hours of telescope exposure to show up in a photograph. Normally the bands of colour in the spectra of stars show what they are made of. But these spectra were unlike anything seen before. Astrophysicists were baffled.

Then in 1963 Dutchman Maarten Schmidt, working at the California Institute of Technology, looked at the spectrum of the radio source 3C 273 and understood what was going on. If a source is old, the wavelength of its emissions will have expanded along with the Universe, making it appear redder (see page 96). However, this radio source was so distant that its colours were reddened out of all recognition; 3C 273 was on the outer fringes of the expanding Universe, 2,000 million light-years away.

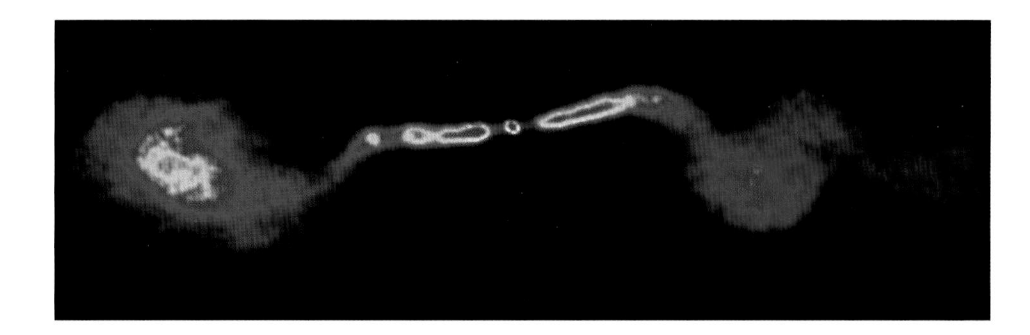

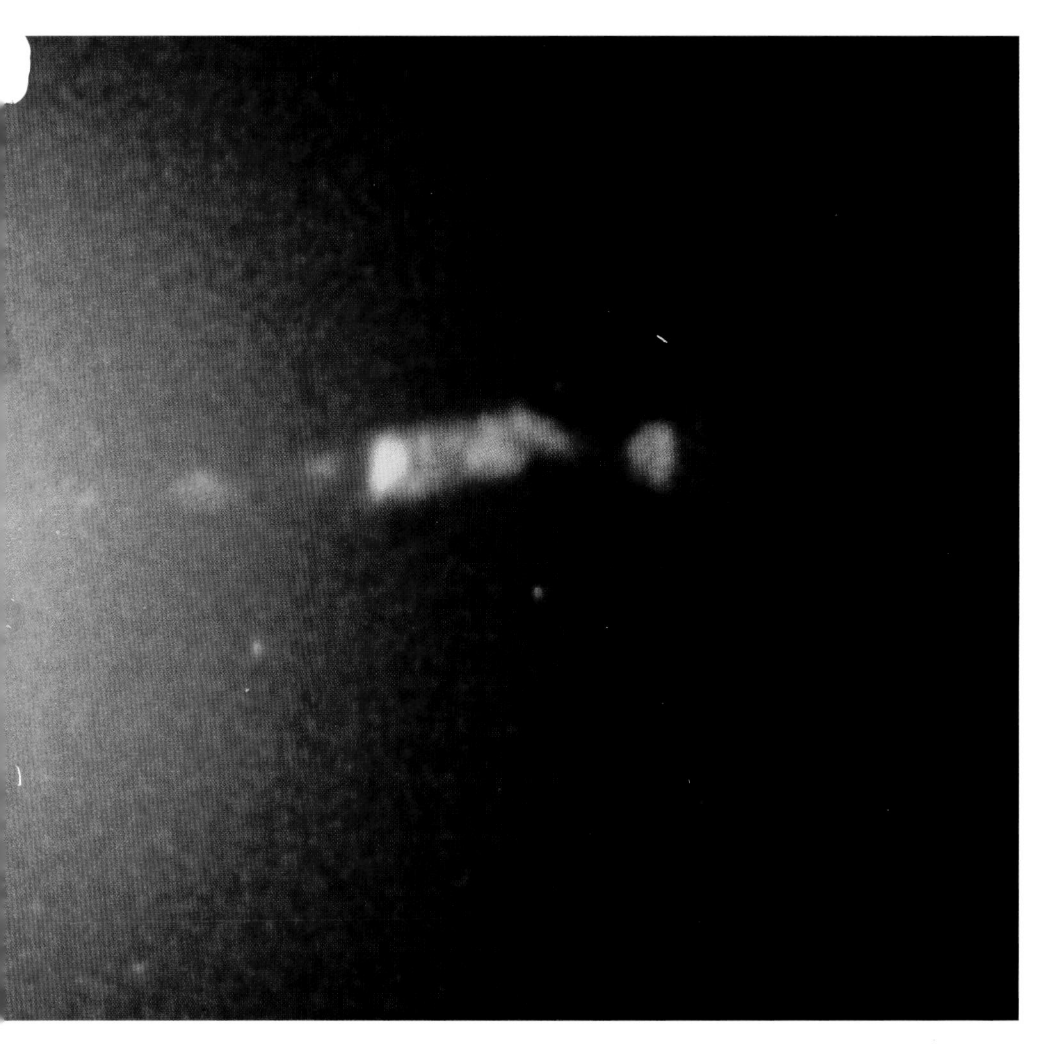

Lobes (top) *The radio galaxy 3C 449 is an elliptical galaxy with a violent core 350 million light-years away. It generates two jets of hot plasma which broaden into radio-emitting lobes.*

Seyfert galaxy (left) *The Seyfert-type spiral galaxy NGC 1068. The bright core has gas moving at speeds up to 5,000 km (3,000 miles) per second. Seyferts may be temporary stages in the evolution of galaxies.*

Cosmic jet (above) *Something strange, probably a black hole with a mass of at least 2 billion Suns, lies at the centre of M87, the nearest active galaxy, a giant elliptical about 50 million light-years from Earth in the constellation Virgo. M87 spews out a glowing jet that extends nearly 5,000 light-years into space. At the core of M87 gas is swirling at 750,000 kilometres (1.2 million miles) per hour. These highly active galaxies are probably in rapid evolution.*

Only a handful of galaxies are that remote. Quasi-stellar radio sources, or "quasars", suddenly hit the headlines.

When astronomers took another look at the radio star 3C 48, they found an even stronger reddening. This quasar turned out to be 10,000 million light-years away. The most remote is PC 1247+3406, discovered by the quasar veteran Schmidt using the Hale telescope in 1991. Light from this far-flung cosmic outpost has taken over 90 per cent of the age of the Universe to reach us.

To be visible at all, such remote objects, beacons in the dawn of cosmic history, have to pump out prodigious amounts of energy. An average quasar, probably no larger than our solar system, is brighter than 100 galaxies, or a million million stars!

GALACTIC POWER

Where can this enormous energy come from? Early speculations centred on chain reactions of supernovae or giant pulsars, but most theorists eventually agreed that quasars need a more powerful engine. This could be a "supermassive" black hole about 100,000 million times heavier than the Sun. Such an awesome object could be created when stars moving towards the centre of a young galaxy get too crowded and collapse into a super-hole. It is possible that particularly powerful quasars have more than one such black hole.

The gravity of the supermassive hole sucks in nearby matter, giving an "accretion disc" as stars swirl round before disappearing down the black hole "drain". This mechanism switches on the quasar, blasting out powerful beams of matter and all kinds of radiation as charged particles are hurled around.

While quasars are still mysterious, there is mounting evidence that they are linked to galactic nuclei and are important in the formation and development of galaxies.

The different kinds of active galaxies – quasars, radio galaxies and Seyferts – may be different stages in galaxy evolution. The first quasar probably appeared when the Universe was less than a 1,000 million years old, when galaxies began to condense.

The dark side of matter

THE MISSING UNIVERSE

A careful look at gravitational motion shows there has to be a lot more to the Universe than meets the eye. With as much as 99 per cent of all galaxies and cosmic material made of invisible "dark matter", we might be a cosmic peculiarity – a freckle on the face of an otherwise very smooth Universe.

In 1932 the astronomer Jan Oort published in a Dutch journal a bizarre idea that he had been mentioning in lectures for several years. Stars and galaxies move under the gravitational pull due to their mass, and from the motion of stars in our galaxy Oort had estimated the total amount of matter that had to be present. However this seemed to be about twice what was visible through telescopes. Perhaps something invisible was out there.

The following year Fritz Zwicky saw the same effect on a larger scale. Measuring velocities inside the Coma constellation, he found that many component galaxies were moving so fast that the cluster would fly apart unless it contained ten times more mass than its light suggested.

This idea of invisible "dark matter" remained a curiosity until Vera Rubin and her colleagues at the Carnegie Institution in Washington DC looked at the way spiral galaxies rotated.

Rubin's group began their study with the Andromeda galaxy, which is 2 million light-years away. Like the Milky Way, Andromeda has an outer spiral, and Rubin expected that stars at the galaxy's outer edge would move more slowly than those near the centre, like the planets round the Sun. She was surprised to find that Andromeda stars had the same velocity, almost regardless of their distance from the galaxy's centre.

At first she thought that Andromeda was a peculiar galaxy, but then other spiral galaxies were found to behave the same way. As evidence accumulated in the 1970s, feeding all the latest data into major computer calculations finally confirmed Oort's suggestion. As much as 90 per cent of spiral galaxies seems to be dark matter, a mysterious outer "halo" that prevents the inner spiral being ripped apart by centrifugal forces.

Dark matter is also essential for cosmology. The Big Bang inflation picture (see page 102) says that the Universe is balanced on a gravitational knife-edge between eternal expansion and ultimate collapse. Visible cosmic mass provides at most only about ten per cent of the critical density needed to provide this knife-edge. The visible Universe was shaped by an immense invisible mould.

HOT AND COLD

The invisible dark matter hidden in the galaxy halos could be just inert matter made of protons and neutrons – "brown dwarf" stars, invisible distant Jupiter-sized planets, intergalactic gas and dust, and black holes. But to get the right amount of light nuclear matter, such as helium and lithium, in the visible Universe, only about ten per cent of dark matter can be accounted for in this way. The rest has to be made of heavy particles that do not affect ordinary nuclear manufacture.

Whatever it is made of, this slow-moving, or "cold", dark matter expands along with the rest of the Universe and its gravity can shape galaxies. But it would have made far too many.

There could also be "hot" dark matter consisting of massless or very light particles zooming about at or close to the speed of light. The natural candidate for hot dark matter is the neutrino, which is invisible anyway. Soon after the Big Bang, floods of neutrinos spilled from the cosmic

VERA RUBIN – ASTRONOMICAL GRANDMOTHER

Vera Rubin announced her first findings concerning the motion of spiral galaxies in 1950 when she was 22 years old, three weeks after her first child was born. She is now 66 and has written a book entitled *My Grandmother is an Astronomer,* which is dedicated to her granddaughter.

Rubin was encouraged to take up astronomy by the famous Big Bang physicist George Gamov. Her findings, which scientists now take very seriously, went more or less unnoticed until 1963. Rubin has spent most of her career at the Carnegie Institution Department of Terrestrial Magnetism and has studied more than 200 galaxies looking for dark matter. She hopes to inspire young people, especially girls, to look for the invisible stuff that makes up most of the Universe.

soup and have travelled aimlessly across space ever since. If each neutrino has a tiny mass, there are enough neutrinos around to supply dark matter.

In an infant Universe full of freely roaming neutrinos, small local irregularities would be washed out quickly, and only large structures, such as superclusters of galaxies, would survive. Smaller objects, such as the galaxies themselves, would have to wait for the break-up of these superclusters, and there has not been enough time for all the galaxies to have broken off. Neutrinos alone cannot have shaped the Universe.

Most cosmologists are now convinced that a blend of hot and cold dark matter is needed to explain the observed structure of the Universe, the former providing the largest scale structure, the latter adding the galactic details.

Most of the cold dark matter must be made of heavy particles which interact only feebly with ordinary matter. There is a long menu for these unseen particles, known as "weakly interacting massive particles", or WIMPS. Supersymmetric particles (see page 84), which could turn up in high-energy laboratory experiments, are high on the list.

Dark cloud (above) *Astronomers using the ROSAT X-ray satellite observatory announced in 1993 the discovery of a huge concentration of dark matter in the small NGC 2300 group of galaxies, 150 million light-years away in the direction of the Cephus constellation. The X-ray pictures show the group immersed in a cloud of hot (10 million degrees) gas about 1.3 million light-years across.*

Waiting for an encounter (right) *In one of the quietest places on Earth, a salt mine 1,100 (3,600 ft) deep at Boulby, on the north-east coast of England, a special detector designed by physicists at the UK Rutherford Appleton Laboratory is waiting for a stray encounter with dark matter particles from outer space. Hundreds of ton(ne)s of water shield the detector.*

In 1993 came evidence for dark matter in the form of "brown dwarfs" – relatively small clouds of gas that have never made it as stars. The only chance of seeing these invisible wisps is when they cross in front of a more distant and brighter star, briefly disturbing the image by making it more intense.

Machos and brown dwarfs

THE FIRST SIGNS OF DARK MATTER?

Seeing the invisible ***The Mount Stromlo Observatory near Canberra, Australia, site of a search for signs of MACHO dark matter that started in 1989. The search has covered about 500,000 stars.***

Our Universe could be full of planet-sized wisps of gas, too small to kindle a nuclear spark and blaze into life as luminous stars. Floating on the outer edge of galaxies, these flimsy wisps could provide some of the missing dark matter of the Universe. Astrophysicists call them "massive astrophysical compact halo objects", or MACHOs. The problem was how to see them if they are invisible.

In 1986 Bohdan Paczynski of Princeton suggested looking for them by focusing telescopes on bright stars beyond the MACHO region and watching to see if something got in the way. As MACHOs cross in front of stars in the nearest galaxy, the Large Magellanic Cloud about 150,000 light-years away, they might disturb the star images. Normally one astronomical object passing in front of another gives an eclipse, temporarily blotting out the distant light, but Paczynski showed that under the right conditions the reverse could happen, the image of the distant star becoming brighter in an "anti-eclipse". "When I proposed this idea, I thought it was science fiction," says Paczynski.

Light is slightly bent by gravity. This was one of Einstein's original predictions, and the pull of the Sun's gravity on starlight was seen in 1919 by comparing the positions of stars during and after a solar eclipse. With light from a distant star being bent both ways round an intervening MACHO, the otherwise invisible object could focus the starlight like a lens, producing a brighter image.

As our galaxy rotates, its MACHO cloud sweeps across the backdrop of distant stars,

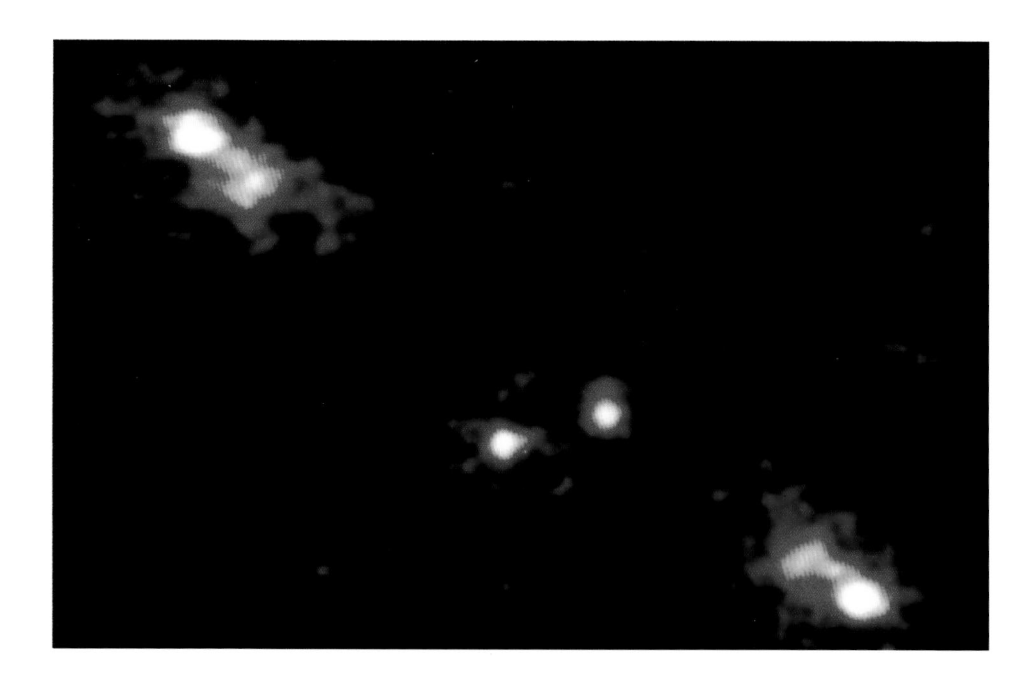

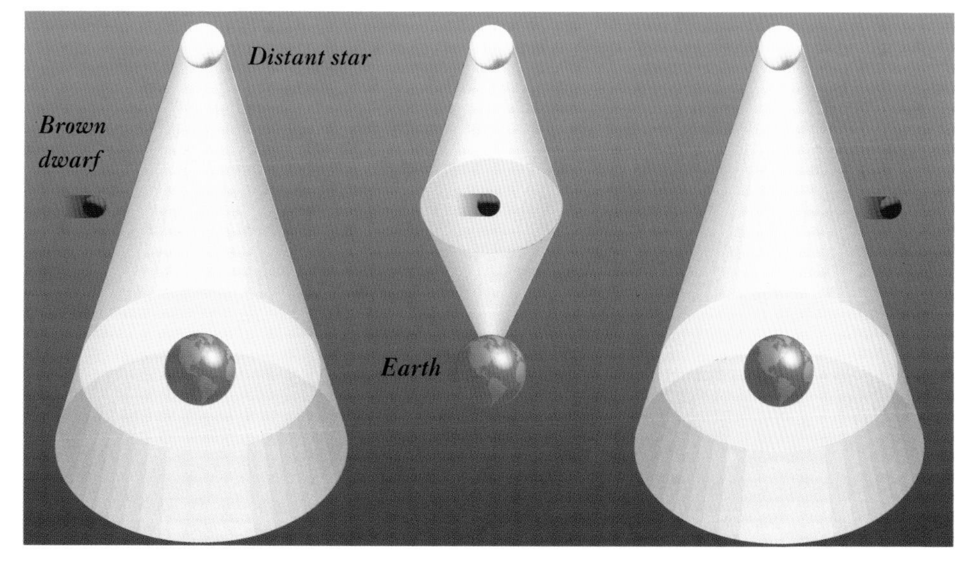

Mirror images (top) *Light from the galaxy AC 114 is bent to create a repeated image by the gravitational force of dark matter in a cluster of galaxies in the foreground of this Hubble Space Telescope image.*

Bending light (above) *Brown dwarfs are made of primeval hydrogen and helium but are too small – about one tenth the mass of the Sun – to ignite thermonuclear fusion and produce light. On the other hand, they still have enough gravity to keep their hydrogen and helium from evaporating into space and will bend light rays from distant stars on their way to Earth.*

like a moving net curtain. Occasionally a MACHO could be in the right place to act as a lens, but after the MACHO had moved across, the focusing would switch off and the star image would revert to its original form. Calculations showed that seeing one such "microlensing" effect would involve looking at about a million stars. It was a formidable task.

TWO TEAMS

In 1989 Charles Alcock at Berkeley suggested a new search for dark matter. When looking for rare phenomena, it is always good to have independent experiments so that initial findings can be quickly verified. Berkeley astrophysicists teamed up with Australian specialists to use the Mount Stromlo Observatory near Canberra in Australia, calling themselves the MACHO collaboration. A French group called EROS (Expérience de Recherche d'Objects Sombres) used a telescope at the European Southern Observatory (ESO) in La Silla, Chile.

The EROS team used two techniques. One approach, sensitive to MACHOs up to about one-tenth the size of the Sun, used giant 5-metre- (50-foot-) square photographic plates, each covering less than one thousandth of the visible sky. From 1989 to 1993, more than 300 of these huge plates were exposed, each for about one hour, at the La Silla telescope. About ten million star images were recorded. The photographic images on the exposed plates were converted into digital data by a special machine at Paris Observatory, scanning in tiny steps a hundredth of a millimetre (a three-thousandth of an inch) across. The total amount of recorded information was equivalent to about a million times that in this book!

The second EROS technique was aimed at smaller MACHOs, at least a thousand times smaller than the Sun. Rather than using photographic plates, this used a chequerboard array of tiny semiconductor chips of a type known as charge-coupled devices (CCDs), such as those that are used to pick up the image in video cameras. The US and Australian MACHO team also used a light-enhancing CCD camera.

In the mass of recorded data, the vast majority of Large Magellanic Cloud star images did not change, but in just a few cases a distant star image suddenly became several times brighter, staying that way for about a month, before reverting to its original level. This temporary focusing corresponds to MACHOs about ten times smaller than the Sun, bundles of potential star material too small to make the grade. However, these "brown dwarfs" cannot be the only constituent of dark matter.

Cosmology goes into space

THE COBE SATELLITE

In 1989 NASA launched its first satellite dedicated to cosmology. The Cosmic Background Explorer (COBE) was designed to look at the faint microwave radiation that permeates space as a relic of the Big Bang and search for clues to the origin of galaxies. This modest spacecraft has been a big success.

Once the microwave background radiation had been discovered (see page 100), astrophysicists wanted to take a good look at it to see if it gave any clues about the Big Bang. Precision measurements are difficult with an effect that is so tiny – the background radiation is 100 million times fainter than the heat surrounding us in everyday life. Undeterred, scientists mounted experiments in balloons, rockets and high-flying aircraft.

In 1974, John Mather, born in New Jersey in 1946 and researching at the Goddard Institute for Space Studies in New York, proposed that NASA build a small spacecraft to study the microwave radiation from above the atmosphere. Already having measured this radiation on Earth, he was thinking of moving on until he saw that NASA was looking for new projects and had put out an "Announcement of Opportunity"

CHEAP SATELLITE

NASA fell for Mather's idea. It was a fairly cheap satellite and the agency therefore had little to lose. Thus the Cosmic Background Explorer (COBE) was born, approved by NASA in 1982. The new spacecraft was ready for launch in 1986, but was postponed when the *Challenger* accident halted the US space programme for four years. In a rush programme, COBE was rebuilt, substantially lighter, for launch by a rocket and finally went into orbit on 18 November 1990.

Two months after launch the COBE team at Goddard, led by Mather, announced the first findings. The Far Infrared Absolute Spectrophotometer (FIRAS), one of the three detectors carried by the satellite, measured the spectrum of the background radiation, and found it fitted exactly that of a perfectly radiating material, or "black body", as required by the simplest version of the Big Bang theory. The temperature of the radiation is 2.735 degrees above absolute zero. The fit is so perfect that 99.97 per cent of the radiation must have been emitted within one year of the Big Bang.

Mather's announcement of this result at the American Astronomical Society in January 1990 brought instant applause. It was a hundred-fold improvement on previous measurements and was a relief to cosmologists. They had been worried when an earlier rocket experiment showed a large deviation from the ideal spectrum.

TOO SMOOTH

New measurements followed, made by COBE's Differential Microwave Radiometer (DMR), producing the first microwave map of the entire sky. It confirmed the cosmologists' picture of a perfectly smooth, uniform Big Bang.

Although this early COBE success earned it a place in science history, a primary mission goal remained. An infant Universe with a Big Bang that exploded equally hard in all directions was too good to be true. Everywhere in space, matter is lumpy, with galaxies of stars, clusters of galaxies and even superclusters of clusters. For these structures to have evolved, there must have been corresponding "microlumps" in the

FIRST LIGHT

The Cosmic Background Explorer, here seen being tested at the Goddard Space Flight Center, is 2.5 metres (8 feet) in diameter and 5.5 metres (18 feet) long, weighing about 5,000 pounds. It circles Earth in a near polar orbit at an altitude of 900 kilometres (560 miles). In winter it can be seen just after sunset and shortly before dawn, pulsating slightly in brightness.

Keeping the COBE instruments at low temperatures to enhance their sensitivity was the major challenge. Two of them were cooled with superfluid liquid helium to minus 271 degrees. DIRBE, the third instrument, looks at the infra-red sky, searching for the first faint light from stars and galaxies as they began to form.

early Universe that left their mark on the microwave background radiation. These primordial density variations were the seeds of galaxies.

In the early 1980s cosmologists had been worried by the continual smoothness of the background radiation, with the temperature the same all over the sky. In 1982 Stephen Hawking pointed out that this was beginning to cramp the eventual outcome of the crucial inflation phase and overconstrain the resulting Universe.

The initial COBE findings underlined this remarkable smoothness, with the temperature the same in all parts of the sky to one part in 25,000. Cosmologists had to polish their calculations to make sure that their estimates of primordial density variations were small enough to have eluded COBE so far. Then in 1992, after carefully checking that all sources of error had been taken into account, the COBE team saw the first tiny signs of texture begin to appear on their maps of the microwave sky. A second brilliant chapter in the life of the amazing COBE satellite was about to unfold – as described on the next pages.

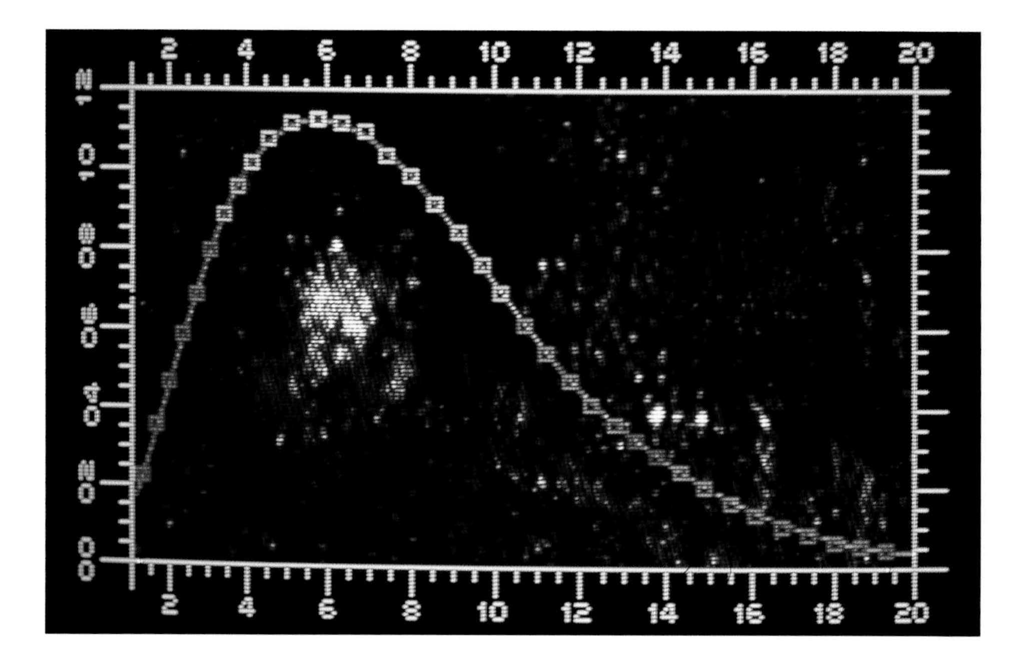

Perfect radiator (below) *The spectrum of the microwave background radiation (the curved line) taken by COBE's Far Infared Absolute Spectrophotometer (FIRAS) instrument. Sixty-seven measurements matched the spectrum of a hypothetical perfect radiator, a "black body", to within a fraction of a per cent, as predicted by the Big Bang picture.*

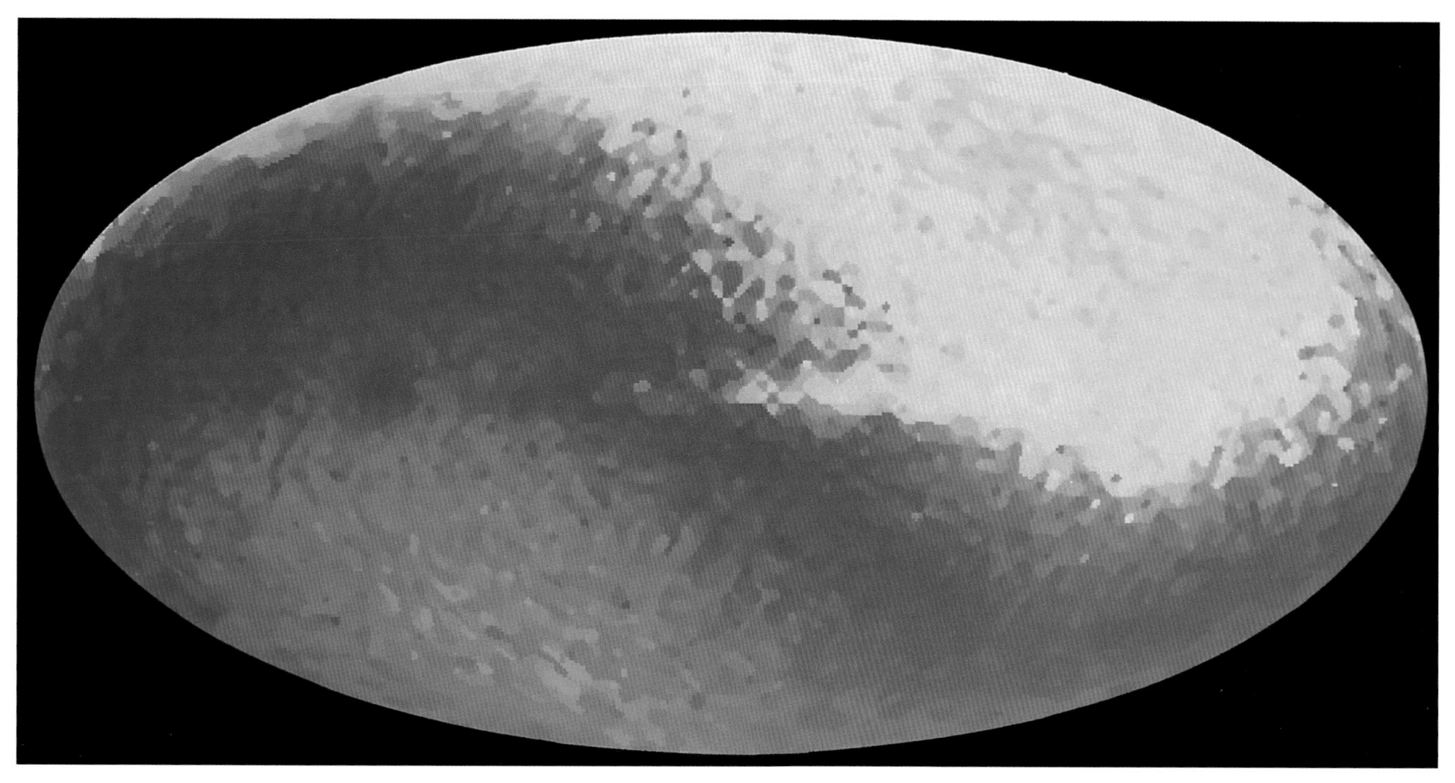

Smooth (bottom) *A microwave map of the entire sky taken by COBE's Differential Microwave Radiometer (DMR). The map shows a gradual blue to pink variation (0.0033 degree), a result of the motion of our solar system through the Universe. When this movement is subtracted, the remaining temperature distribution is very smooth.*

Seeds of the Universe

RIPPLES FROM THE BEGINNING OF TIME

The Big Bang's tiny space-time bubble was full of quantum shimmer. When the mighty force of inflation was unleashed, this shimmer became a lacework, indelibly imprinted on the fabric of the Universe. Here was the stencil that guided gravity's mighty brush as it painted the cosmic backcloth.

The space and time created by the Big Bang was not empty. Confined to subatomic dimensions, the first tiny fraction of a second was wild and chaotic, ruled by the uncertainty principle. Particle–antiparticle pairs and other quantum clutter continually flashed on and off, making the vacuum glitter with tiny bursts of transient energy.

But 10^{-35} of a second after the Big Bang, the infant Universe-dot was gripped by the powerful inflation force as energy was released into the vacuum, ballooning the initial space-time bubble out faster than the speed of light (see page 102). The random quantum blips were dragged along. Trapped in the mighty amplifier of inflation, they exploded outwards, reprieved from their quantum fate of vanishing forever.

At 10^{-32} second, when inflation abruptly stopped, the Universe that we can now see was tennis-ball sized. Filled with inflated blips, it was no longer smooth. The seeds of the Universe had been sown. It was now up to gravity to make them grow. Gravity is a one-way force,

MAKING RIPPLES

George Smoot of Lawrence Berkeley Laboratory leads the science team that operates COBE's Differential Microwave Radiometer (DMR), the precision instrument that scanned the cosmic background radiation and discovered the cosmic "ripples". These tiny fluctuations, one of the biggest scientific discoveries of the century, made George Smoot a science celebrity. Born in Florida in 1945 and the son of a geologist, Smoot is a member of a committee at Berkeley that seeks to give a better public appreciation of science.

Ripples in the background *COBE's tiny temperature variations seen in 1992 – only 30 millionths of a degree – were the seeds that gravity nurtured into a Universe. The idea that small irregularities could clump together under gravity was first suggested by Isaac Newton more than three hundred*

always pulling matter together. Gradually more and more material snowballed, laying the foundations for galaxies.

In the early 1970s, the British theorist Ted Harrison and the Soviet cosmologist Yakov Zeldovich showed that all the original quantum blips would have undergone the same expansion. As when inflating a balloon covered in dots, the quantum blips got further apart but the relative pattern remained the same. The pattern of gravitational force would have also been preserved on the cosmic background radiation.

On 23 April 1992 at a meeting of the American Physical Society in Washington, DC, astrophysicist George Smoot from the Lawrence Berkeley Laboratory near San Francisco described to an astonished audience how NASA's COBE satellite had seen patches of hot and cold in the sky, tiny temperature ripples of only 30 millionths of a degree. Rarely has a scientific discovery been so widely and readily acclaimed. The next day the press all round the world carried banner headlines, such as "How the Universe was made". "It is the scientific discovery of the century, if not of all time", said Stephen Hawking. George Smoot when presenting the results made the much quoted comment: "If you're a religious person, it's like seeing the face of God."

Eight months after COBE's astonishing revelation, independent measurements of the microwave background radiation by a team of scientists from the Massachusetts Institute of Technology, NASA and Princeton confirmed the existence of micro-ripples in the background radiation.

Although confined to a high-altitude balloon at 40 kilometres (25 miles) rather than a satellite lofted to 900 kilometres (560 miles), the detector was 25 times more sensitive than COBE's and a six hour exposure was enough to reveal the tiny ripples. The similarity of the tiny fluctuations measured under different conditions by the two precision experiments confirms the Universe's quantum origin.

With COBE's scientific mission now over, the next task is to match the ripples to today's Universe.

years ago. If matter were dispersed throughout infinite space, some of it would convene into one mass and some into another to make an infinite number of greater masses, scattered great distances from one another throughout that infinite space. And thus might the Sun and the fixed stars be formed.

COBE IN PERSON

John Mather of NASA's Goddard Space Flight Center had the idea for the COBE satellite and led the science team that first measured the microwave spectrum. He is also largely responsible for pushing the project through despite the set-backs in the aftermath of the Challenger disaster. To many he is COBE in person. Born in New Jersey in 1946 and the son of a farmer, he was was fascinated early in life by astronomy and telescope-making. "What startles me the most is that we really can make a coherent story of the Big Bang even though it happened so long ago in such extreme conditions," he says.

How old is the Universe?

MEASURING HOW SPACE EXPANDS

The difficulties of measuring the distances to galaxies are at the root of a long battle over how fast the Universe is expanding. These uncertainties put doubts on the age of the cosmos. Some of the estimates absurdly imply that the Universe is younger than its oldest stars!

In 1929 Edwin Hubble discovered that the Universe is expanding – distant galaxies appeared to be rushing away at velocities proportional to their distance from the Milky Way. The further away a galaxy is, the faster it appears to recede (see page 100). This apparent velocity-to-distance proportionality is known as the Hubble ratio.

Actually the distant galaxies are not themselves receding. It is just that the Universe is getting bigger, with the intervening space still expanding in the wake of the Big Bang. The Hubble ratio tells us how fast this happens and is gives a clue as to how old the Universe is. The faster the expansion, the less time has passed since the Big Bang. The gravitational pull between all the matter in the Universe gradually slows down the expansion, and to arrive at the age of the Universe, the Hubble ratio has to be put into an equation with the density of mass in the cosmos, a quantity that is notoriously difficult to measure and which nobody knows for sure.

Edwin Hubble initially put the ratio at 530, a gross overestimate caused by errors in his measurements, implying that the Universe is only one or two thousand million years old. The work was continued by Allan Sandage, who in 1956 arrived at a value of 180, corresponding to a Universe some 5,000 million years old. Later Sandage collaborated with the Swiss astronomer Gustav Tammann, bringing the value down to 50 and giving an age of some 12,000 million years. However, these age estimates depend on the highly uncertain cosmic mass density.

In 1976 French astronomer Gerard de Vaucouleurs of the University of Texas challenged Sandage's value, arriving at a Hubble ratio of 100 and a 10,000-million-year-old Universe. There has been a long and acrimonious battle between the two camps, both claiming they were right, each with carefully estimated errors which exclude the other's result! One possibility is that the ratio is not just a fixed number, but could have different values for various stages in the Universe's evolution.

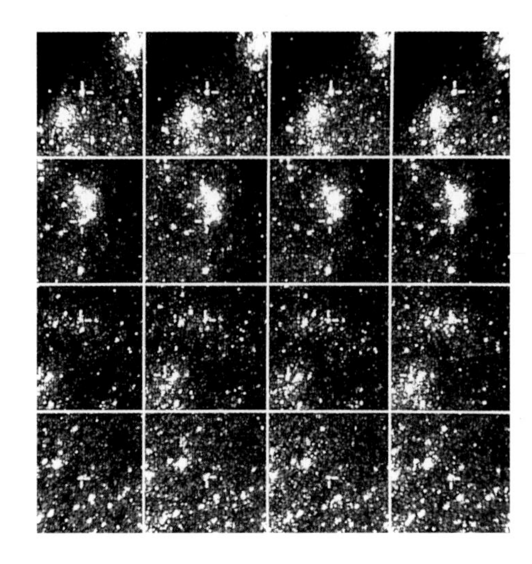

Cepheids in M81 ***The Hubble Space Telescope found more than 30 Cepheids (indicated by white lines in each picture) in the spiral galaxy M81. Previously only two Cepheids had been identified in the galaxy. The Hubble discovery has given the most accurate distance estimate for M81 yet: 11 million light years. Cepheids are yellow supergiant stars that pulsate regularly, swelling and brightening with periods ranging from 1 to 20 days. Cepheids provide a standard against which other stars can be measured. The longer the pulsation period, the more luminous the star.***

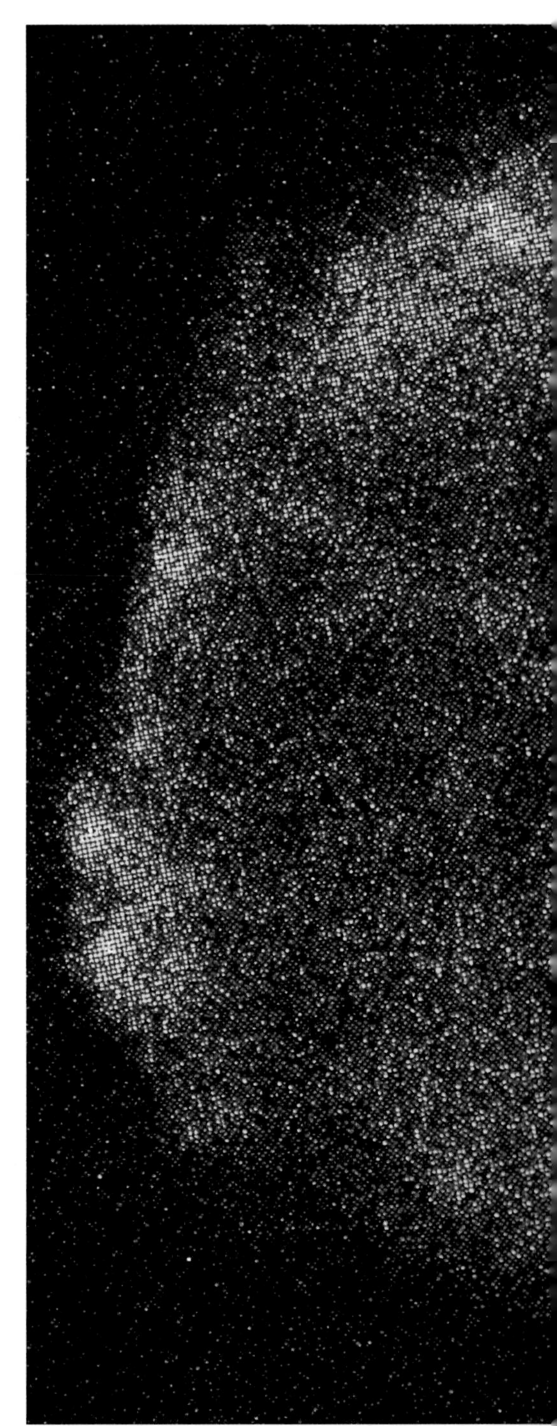

Tycho's star *The remnant of the Type I supernova, SN1572, that exploded in the constellation, Cassiopeia in 1572. The supernova, visible in daylight, was observed by the Danish astronomer Tycho Brahe. The gas from the explosion, still expanding at 11,000 km (6,800 miles) per second, is here imaged by the X-ray satellite ROSAT.*

THE AGE PARADOX

Most cosmologists believe the Universe is 15,000 million years old. Stellar astronomers, however, claim that the oldest stars, found in densely packed "globular clusters", are 16,000 million to 19,000 million years old. This raises the "age paradox". The oldest stars clearly cannot be older than the world in which they live. Either the age of the Universe has been miscalculated or there is something "wrong" with the globular cluster data.

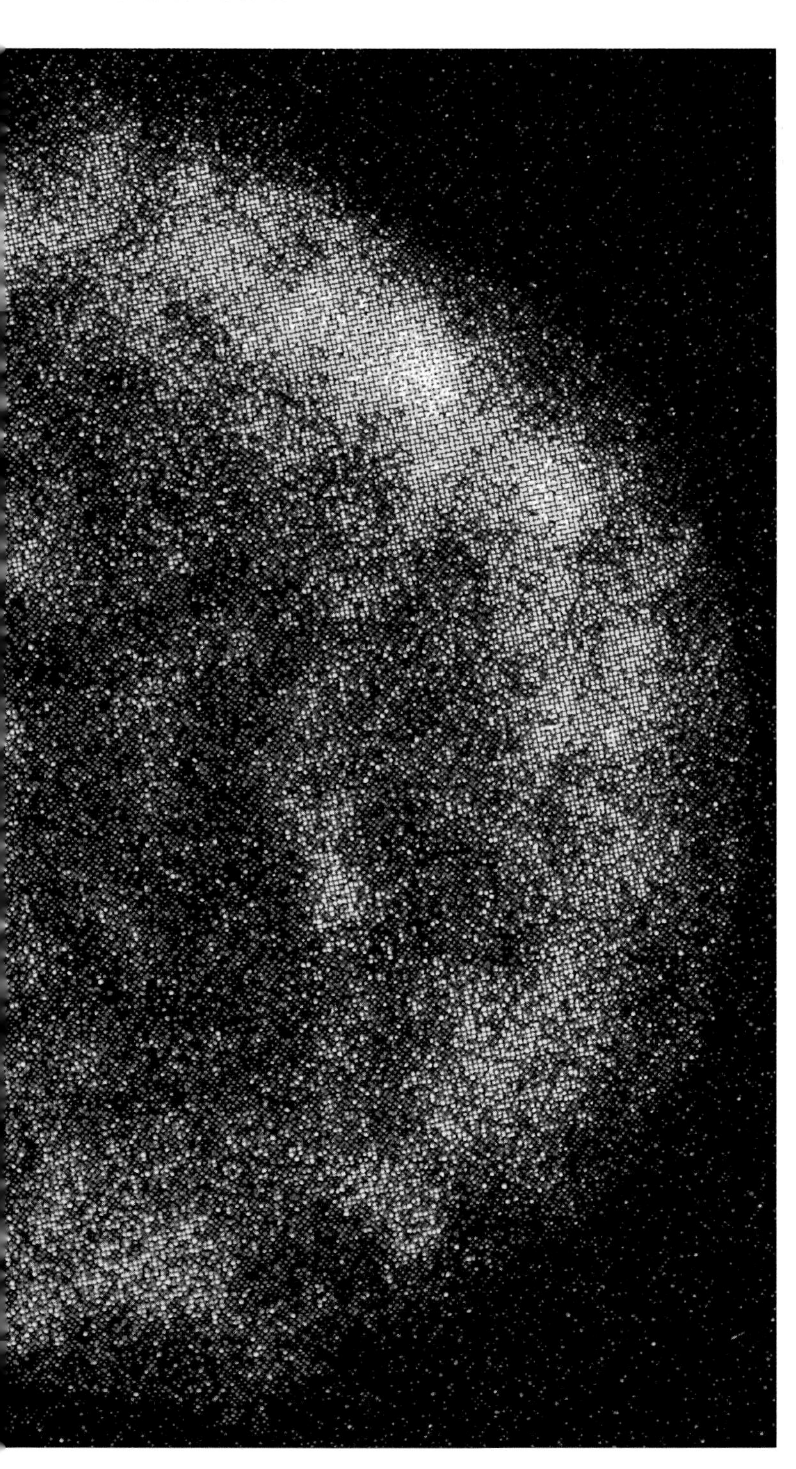

HUBBLE FLOW

The first part of the Hubble ratio equation is determined by measuring the "red shift" (see page 97), the apparent reddening of light from distant galaxies caused by the cosmic expansion or "Hubble flow". With the movement of nearby galaxies dominated by strong close-quarter gravitational grip, astronomers have to look as far out as the Virgo Cluster, 50 million light-years away, or beyond to interpret the red shift. Here galactic motion is dominated by the expansion.

The other task is to fix galactic distances. This is the difficult bit. Several methods are applied, depending on how far away the objects are. For relatively nearby galaxies the traditional way is to use variable stars called Cepheids, discovered by Henrietta Leavitt in 1912. However, such stars, pulsating regularly as they expand and contract, can only be used as reliable distance indicators out to about 25 million light-years.

Recently, planetary nebulae have proved an alternative and reliable indicator for nearby galaxies.

COSMIC CANDLES

So-called Type Ia supernovae now look like an increasingly promising standard "candles" for distance measurements beyond 50 million light years. Such supernovae are usually to be found in elliptical galaxies or older spirals and occur when a white dwarf star accumulates material from an orbiting neighbour star. Suddenly reaching a critical mass (the Chandresekhar limit) the dwarf star explodes.

Because they are produced in the same way, these supernovae all light up with the same brightness. They are more luminous than the more commonly known Type II supernovae (see page 108). Although measuring the peak brightness of white dwarf explosions is difficult, this method could extend the cosmic "ruler" out to 1,000 million light years or more.

A nearby Type Ia supernova was observed in 1937 in the galaxy IC4182, 16 million light-years away. Using the Hubble Space Telescope a team of astronomers, including Allan Sandage, in June 1992 turned to this galaxy to look for Cepheids and obtain a better distance calibration. They found 27 Cepheids and arrived at a Hubble value of 45, consistent with a Universe 15,000 million years old, the generally agreed estimate. The battle over the ratio, however, has continued.

The ultimate quiz game

PHYSICS AND METAPHYSICS

During the twentieth century, progress in understanding has led to a profound awareness of humanity's precarious foothold on a vast ladder. Perched on a fragile rung, peering down towards atoms and gazing up towards stars, we have only just begun to fathom the cosmic order.

"Fear and the blindness of thought cannot be chased away by the brilliant light of the Sun – only through a study of Nature and the laws that govern it," wrote the Roman poet and philosopher Lucretius in his first-century BC book *De Rerum Natura*, which popularized the Greek atomic theory. These first atomists aimed to discredit belief and superstition, formulating a comprehensive natural explanation of the world with no call for the supernatural.

The twentieth-century discoveries of the structure of the microworld demonstrated an inherent complexity undreamed of by earlier generations. This search to discover the ultimate bedrock structure of matter led physicists deeper and deeper into the microworld, entering the unfamiliar quantum domain where everyday ideas such as cause and effect take on new meanings.

This journey into the microworld went hand in hand with a demand for increasingly higher energy "microscopes" to illuminate the way. These higher energies took physicists back in time, recreating conditions from the primordial Universe and pointing to the Big Bang – when a Universe larger than we can see and more complex than we can imagine was created in a single pinpoint explosion about 15,000 million years ago.

Now, some 2,000 years after Lucretius was writing, science could have achieved the ultimate aim of those early atomists. It now seems possible to explain how the Universe began without requiring a deity to "push the button". Under the laws of physics, a universe can suddenly appear as a random quantum event and create itself out of the void.

THE DEMISE OF ARROGANCE

Five hundred years ago, most people arrogantly maintained that the Earth was the centre of the Universe, but Galileo showed that the Earth revolved around the Sun. Then came the realization that the Sun was in a fairly remote corner of our galaxy. Far from being the glory of the heavens, the Sun is just a modest star among millions of others. The next reappraisal came early in the twentieth century, when the research of Edwin

The scientific method (above and right)
The search for new answers is symbolized here by the Mauna Kea Observatory, Hawaii (above), and an oxygen ion collision photographed at CERN (right). The tool of science is the scientific method, the careful interplay of intelligent premise and thorough experiment. Over the centuries, systematic and precisely controlled measurement has become increasingly important. Searching for new clues taxes imagination and ingenuity to the limit.

THE DIM BOUNDARY

"The exploration of space ends on a note of uncertainty. And necessarily so. We are, by definition, in the very center of the observable region. We know our immediate neighbourhood rather intimately. With increasing distance, our knowledge fades, and fades rapidly. Eventually we reach the dim boundary – the utmost limits of our telescopes. There we measure shadows, and we search among ghostly errors for landmarks that are scarcely more substantial." *Edwin Hubble,* The Realm of the Nebula.

Hubble and others showed that the Milky Way is not alone, that the Universe is made up of some 100,000 million widely scattered galaxies.

Aware that the Universe has to contain large amounts of invisible dark matter, astrophysicists suspect that most of the Universe could be made in an unfamiliar way. Nuclear material as we understand it, the stuff that our world is made of, was only a cosmic afterthought, formed late on in the aftermath of the Big Bang. By the time this nuclear die was cast, most of the rest of the Universe could have already been moulded.

Instead of arrogance, we now have humility. While we can catalogue more and more understanding of the world immediately around us, we do not know what, or even where, most of the Universe is! Nature seems to be playing some perpetual quiz game where, each time we get the answer right, a new mystery jackpot appears and the stakes are raised.

WHAT'S THE POINT?

The more we understand about the Universe, the more pointless it seems, suggested Steven Weinberg in his acclaimed book *The First Three Minutes*. Weinberg studied philosophy before he went into physics, and is a leading scientific thinker advocating an ultimate unified theory.

Other scientists are more cautious. "The Universe is so mysterious in being tuned the way it is that I am willing to believe there is a purpose", said Allan Sandage, one of 27 cosmologists who were confronted with the pointless Universe in the book *Origins* by Allan Lightman and Robert Brewer. "Weinberg understands more and more of the Universe. I understand less and less." said Maarten Schmidt, the discoverer of quasars.

BIG FRAME, SMALL PICTURE

Trend-setting particle physicists and cosmologists have a flair for finding compelling names like a "theory of everything", the popular name for the ultimate unified theory. But it would be misleading to suggest that science is now filling in the final blanks of the ultimate questionnaire. Big puzzles like "Where do the laws of nature come from?" and "Why does the Universe bother to exist?" remain.

The scoreboard on the nature quiz game records a high mark for the detection by the COBE satellite of cosmic "ripples". Left for the next round of the quiz are finding the exact value of the Hubble ratio and determining the fate of the Universe – whether it will expand or contract. Finding supersymmetric particles, the prime candidates for dark matter, would push the stakes even higher.

While today's science understands more and more, the overall framework of the picture has unfolded even faster, continually pointing to more that is not understood. Faced with this dilemma, scientists have become increasingly humble. Recognizing the difficulties of finding absolute answers, they diligently continue their "Search for Infinity".

Index

PICTURE ACKNOWLEDGEMENTS

AIP Niels Bohr Laboratory 108, /Dorothy Crawford Collection 130, /Dorothy Davis Locanthi 99, /Harvey Pasadena 61 top, /Physics Today Collection 101, 107 top; Bettmann /UPI 52, 65, 67 top; Bridgeman Art Library /Gavin Graham Gallery, London 14–15; Brookhaven National Laboratory 58, 61 bottom, 70 left, 85 top, 113 bottom; University of Cambridge Cavendish Laboratory 39; CERN 57 top, 59 top, 63 left, 69, 73 top and bottom, 74, 75 top and bottom, 78, 79, 82, 83 top and bottom; Jean-Loup Charmet 22, 28, 92, 95 top; ESA 114; Mary Evans Picture Library 16, 17 top, 46 top; Fermilab Visual Media Services 86; Harvard–Smithsonian Center for Astrophysics 126 top; Harvard University /Joe Wrinn 126 bottom; Hulton Deutsch Collection 19 top right, 20 top, 25, 37 top, 49, 96; Interfoto /Archiv 30 bottom, /Karger–Decker 36 bottom; Joint Institute for Nuclear Research /Yu. Tumanov 115 top; Lawrence Berkeley Laboratory 136 bottom left; Los Alamos National Laboratory 47 top, 113 top; MIT /Donna Coveney 102; Roger M. Macklis 26; Manchester City Council 20 bottom; Mount Stromlo and Siding Spring Observatories 132; NASA 97 bottom, 112, 118, 120, 121 top, centre right and bottom left, 122–23, 124, 127 top and bottom, 128–29, 131 top, 133 top, 134, 138 bottom left; Reed International Books Ltd /Michael J.H. Taylor 53 top; Roger Ressmeyer, Starlight 100; Rex Features /Sipa 85 bottom; Ann Ronan/Image Select 18 bottom, 21 top, 23, 24 bottom, 29 bottom, 94, 95 bottom; Rutherford Appleton Laboratory 131 bottom; The Science Museum 17 bottom, 21 bottom, 31 bottom, 37 bottom; Science Photo Library 8–9, 24 top, 27 top, 34 bottom, 48 top, 93 bottom, /CERN 68, 141, /Fred Espenak 93 top, /Hale Observatories 97 top, /David Hardy 117, /Anthony Howarth 119 bottom left, /Mehau Kulyk 135 top, /Lawrence Berkeley Laboratory 51, 54, 55, /Dr Jean Lorre 8, 128 left, /NASA 122 bottom left, 123 bottom right, 125 top and bottom, 135 bottom, 136–37, /National Library of Medicine 15 top right, /NOAO 107 bottom, 109, /Novosti 56–57, /NRAO, AUI 129 top, /David Parker 71 bottom, 84, /Max Planck Institut für Physik und Astrophysik 138–39, /Roger Ressmeyer, Starlight 90–91, 140, /J.C. Revy 27 bottom, /John Sanford 106, /Robin Scagell 111 right, /Dr Rudolf Schild 115 bottom, /Science Source 66, /Smithsonian Institution 111 left, /U.S. Army 41; Stanford Linear Accelerator Center 44, 62; Sygma /Abe Frajndlich 119 bottom right; Topham Picture Library 33; Pedro Waloskuk 87; Andy Warhol Foundation 45 left and right.

Every effort has been made by the publishers to credit organizations and individuals with regard to the supply of photographs and illustrations. The publishers apologize for any omissions which will be corrected in future editions.

ARTWORK

Julian Baum 12–13, 34–5, 42–3, 76–7 right-hand rows, 80–81 top, middle and bottom, 89, 98–9, 104–5, 110

Keith Williams 14, 19, 22–3, 29, 30, 31, 32 top and bottom, 36, 38, 40, 46–7, 48, 50, 53, 55, 56, 58–9, 60, 63, 64, 67, 69, 70–71, 72, 76 left two rows, 77 top, 88, 97, 102, 103, 116, 119